FLOWER FLIES OF MINNESOTA

FLOWER FLIES OF MINNESOTA

SCOTT KING

Pollination Press LLC
MINNESOTA

DEDICATION

I.m. Phyllis Neese and Arne Petrich, parents of whom I have no memories.
And to Carol, Susan, and David, my newly-met siblings.

Flower Flies of Minnesota

Published by
Pollination Press LLC
Minnetonka, MN 55345
www.pollinationpress.com

First edition.

Printed in the United States.

ISBN 9780991356324

Keywords: flies, flower flies, hover flies, syrphid flies, Syrphidae, insects, beneficial insects, pollination, pollinators, biodiversity, ecology, ecosystem services, ecosystem management, landscape, biology, science, entomology

TABLE OF CONTENTS

TABLE OF CONTENTS

TABLE OF CONTENTS

ACKNOWLEDGMENTS

This book is built upon the work of many others. From the taxonomists who first collected and described the species to the growing legions of citizen scientists contributing photographs to online databases, the contributors to the knowledge of flower flies in Minnesota are almost numberless. Nonetheless, I'd like to enumerate those individuals and organizations that have assisted me in the endeavor of learning and compiling the information contained in this book.

It has been a challenge to learn to identify flower flies. What progress I've made would not have been possible without the yeoman-like efforts of the many experts at bugguide.com and the revelatory online, e-environment created by the people at iNaturalist and the University of California that supports that platform. Countless hours were logged at both of these sites, comparing photos, looking at maps.

I'd like to thank the following people for their assistance and support: Merle Fossum for enthusiastically sharing his property, allowing me to visit the replanted prairie, the woods, and the bog. Dale and Therese Jacobson, poet and artist, for hosting me on a number of visits, humoring my various pursuits of dragonflies and flower flies—I'm grateful for the friendship over the years, your fine flower garden, and the exciting games of chess. Freya Manfred and Tom Pope, extraordinary listeners and empathetic speakers. The dragonfly folk: Kurt Mead, Mitch Haag, Curt Oien, Ami Thompson, Robert DuBois, and Ken Tennessen for many many memorable days afield. Kurt and Betsey Mead for their hospitality and joie de vivre; it was in their vegetable garden Ceriana was photographed. Mary Lysne for a day on the Canadian side of Rainy Lake; Parhelophilus rex was photographed that day. Gretchen Mehmel and family for their hospitality at Norris Camp. Heather Holm, without her enthusiasm for pollinators and native plants and without her talents with graphic design, this book would not have existed. My virtual friends and colleagues on iNaturalist: including Paul Bedell, Nick Block, Even Dankowicz, Brian Gooding, Denis Doucet, Sam Kieschnick, John Klymko, Greg Lasley, Janet Nelson, Erland R. Nielsen, Greg Seitz, Caleb Scholtens, Katja Schulz, Joe Walewski, and Gary Walton. Jeff Skevington and Martin Hauser for removing a number of question marks from the text. Brad Boldan for organizing an annual bioblitz in southern Minnesota. Ron Huber, curator of the insect collection at the Science Museum of Minnesota, for the invitation to visit and work with the collection. James Lenfestey, poet and supporter of good causes, for his boundless energy and persistent work to raise awareness of climate change. John Luhman for his efforts at expertly curating the Syrphidae at the University of Minnesota Entomology Collection. Robin Thomson, curator of the University of Minnesota Entomology Collection. Sam Demas for the conversations and rambles, one of which led us to an encounter with Sand's Fen Fly. The colleges of Carleton and St. Olaf for their concern for and care of important natural areas. A shout out to Diane Angell and Nancy Braker. The city of Northfield for declaring itself a pollinator-friendly city. Xerxes Society for its ongoing dedication to the preservation of invertebrates. Karin Jokela for her efforts to protect pollinators and promote beneficial insects among Minnesota farmers. To Joe and Nancy Paddock, to Edith and John Rylander, all protectors of this planet's earthy delights, all people I greatly admire. And Lisa and Lida, whom I love dearly, this book would never have been begun or finished without their unquestioning support.

A surprising number of people (one could venture to say most people) have never heard of a flower fly. I came to appreciate this fact rather acutely when I tried to explain to people that I was working on a guidebook about flower flies and was met with blank stares and raised eyebrows. My hope is that this book will help in this regard. Finally, I ask for leniency in judgment, agreeing with Williston who wrote (1886) "My work must have shortcomings and errors; the subject is by no means an easy one. I only ask that it be accepted as an honest endeavor to advance the study of these neglected but interesting insects."

HOW TO USE THIS GUIDE

Grass Skimmers

The Grass Skimmers (***Paragus (Paragus)*** spp.) are tiny, Nearctic flower flies. The scutellum is edged with yellow. The face is yellow with or without a black, medial stripe. Eyes have several stripes of pile. Dissection is required to separate species.

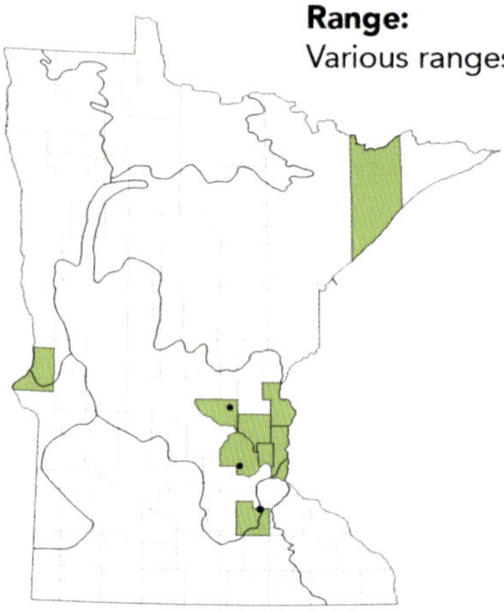

Range: Various ranges

Field Notes: Common, but easily missed because of size. Look for these flower flies perched on low vegetation at the edge of open areas and trails.

Similar species: Most similar to other *Paragus* species. Could be mistaken for *Eumerus*.

Status: Common

Flowers: While often found perched at the tip of grass blades and other low vegetation, they have been observed visiting white snakeroot in Minnesota.

Flight Dates: May - September

Range

This indicates where the main geographical range for a species is located. Because Minnesota is centrally located in North America, this corresponds also to the direction from the state to the center of individual ranges. Thus if a species has an eastern range not only is it's main distribution the eastern part of the continent but it is also east of Minnesota.

Every effort has been made to ensure that this guide is as comprehensive as possible. Not every species has been included in the species accounts, however, omitted species are noted. And all currently known species along with a number of potential species are included in the checklist at the end of the guide.

Maps

Shaded counties represent historic data, data reported in scientific publications and data gathered from museum specimens. Typically this represents observations prior to the year 2000. Points on the map represent modern data, data submitted to bugguide.com and iNaturalist. The dark lines are level III ecoregions. The light lines are county boundaries.

Status

Abundant if greater than 100 observations; Common if the number of observations between 20 and 100; Uncommon between 6 and 20; Rare if less than 6 observations.

This guidebook crosses a boundary and can lead you across as well. Unlike the pursuit of birds or butterflies, creatures large enough to notice and observe without the aid of special knowledge or special equipment, the pursuit of flower flies leads one into the world of mimicry and optical crypsis, into the world of complex ecological interactions, into the world of fugitive species, species both wary and small. These are fascinating paths. An effort needs to be made, active searching undertaken, to find many of the flower flies, though it is true that an elegant number will show themselves at flowers outside your front door.

Chapter One
INTRODUCTION

Introduction

WHAT ARE FLOWER FLIES?

Flower flies are insects. And like most insects, flower flies are beneficial, even essential. And yet very few people are familiar with these particular flies. Probably because there are simply so many kinds of insects! The conspicuous insects continue to be the most well known: the butterflies and moths, the dragonflies, the bees and wasps and ants, the beetles. These garner the most attention, the celebrities of the arthropod world, the subjects of numerous guidebooks. Flies, however, land much lower on any such list. And when they receive any recognition it's almost always as a nuisance, rarely for their looks or benefits: for instance, the most widely recognized flies include the blood-thirsty mosquitoes and deer flies, the annoying house flies. Hopefully, a knowledge of flower flies will change your outlook on flies.

EVOLUTIONARY HISTORY

Flower flies—often referred to as hover flies, pollen flies, or syrphid flies—originated alongside flowering plants some 90 to 145 million years ago and were most likely among the very first pollinators on the planet (long before bees). The dates of origin are not exact and are based upon better evolutionary knowledge of closely related fly families. Direct evidence of their age, based upon the fossil record, remains sparse, with the majority of early flower fly fossils found in Baltic amber, dating to 23 to 52 million years ago.

Flower flies represent but a small fraction of an immensely large group of insects, the order Diptera, which includes mosquitoes, gnats, crane flies, robber flies, house flies, and many other kinds of flies. The name "Diptera" means "two-wings," encapsulating the main characteristic of this order of insects. Most other insects—such as Hymenoptera (bees and wasps), Lepidoptera (butterflies and moths), or Odonata (damselflies and dragonflies)—have four wings. The order Diptera is expansive, including 188 families of flies and something like 150,000 described species and countless undescribed species. Flower flies belong to just one of these 188 families, the family Syrphidae.

POLLINATION BY FLOWER FLIES

The family of flower flies (Diptera: Syrphidae) radiated and diversified into a remarkable variety of forms and colors. This diverse family of flies, while they do not collect pollen as do bees, share a dependence upon pollen as adults. And, like bees, they transfer pollen from flower to flower, aiding in the pollination of the plants they visit. Interestingly, many of these flies are remarkable mimics of the wasps and bees that share the flowers with them. In addition, as will be discussed later in the introduction, the larvae of flower flies perform fascinating and important roles in ecosystems.

IMPORTANCE OF FLOWER FLIES

Flower flies contribute to a great variety of habitats and ecosystems; each species is important in their own small yet complicated way. Some of the flies are specialists: for instance, flies of the genus *Microdon* live only in ant colonies and flies of the genus *Brachyopa* live only in sap leaking from living trees. Many other flies are generalists, especially as adults: for instance, flies of the genus *Eristalis* live pretty much anywhere there is water and decaying organic matter (which is pretty much anywhere) and flies of the genus *Eupeodes* can survive anywhere there are aphids and aphids of any variety.

Mainly, the flower flies are recognized for their role as pollinators. Because flower flies visit flowers and feed on pollen and nectar, they transfer pollen from flower to flower and act as important pollinators. A recent study in New Zealand found some of the flies to be just as efficient at pollination of crop plants as honey bees.

Flower flies visit native flowering plants and garden flowers from spring to fall throughout Minnesota. Early-blooming trees, the first prairie pasqueflowers, and the sudden carpet of woodland ephemerals all attract spring species. Goldenrods and asters attract fall species. And in the months between, these fascinating flies will be found at and around the whole summer succession of blooms.

However, it's not all goldenrod, wild plum, and roses. In addition to their beneficial role as pollinators, the flower flies have additional, and no less important, roles in nature and gardens. The larvae of many common species prey on aphids, controlling the populations of these potential garden pests. Many other common species feed on decaying plant matter. A few of the weirder species prey upon the larvae of ants and social wasps and bees.

Some of the larvae live in truly squalid conditions. Those that thrive, for instance, in anaerobic decaying sludge like that found in waste treatment ponds and feedlots. These specialists have evolved long, breathing tubes, allowing them to siphon in oxygen from the surface. These larvae are commonly referred to as "rat-tailed maggots." Some of these same flies have been reared from the deliquescing organic matter of carrion. It has been suggested that the "bees" Samson finds swarming the carcass of the lion (Judges 14: 5-9) may have been bee-mimic drone flies. As mentioned above, it's not all roses. Nonetheless, these flies carry out important work in our ecosystems, filling an important ecosystem role as recyclers of organic matter.

FLOWER FLY DIVERSITY

The family Syrphidae includes more than 6,200 described species of flower flies worldwide. More than 800 species can be found in the Nearctic. The recent publication for northeastern North America includes coverage of 416 species (Skevington et al, 2019). Probably close to 200 different species of flower flies will eventually be found in Minnesota. The current state list includes 184 recorded species (with an additional 53 hypothetical species). Of those, 138 species are covered in this guide.

Big-headed Fly (Pipunculidae)

FLOWER FLY TAXONOMY

Taxonomy for any insect family can be quite complicated and fluid, and the taxonomy of the Syrphidae is no different. Many scientific names used in early publications are no longer in use. Thankfully, flower fly classification has begun to stabilize, especially as more and more genetic work is performed. Previous synonyms, especially those names used in publications relevant to the Minnesota species, have been included in the checklist at the back of the guide.

One long-standing question has been where to place the pipizines? This guide adopts the subfamily Pipizinae as a sister subfamily to Syrphinae. The phylogenetics of syrphidae, as currently understood and as used in this guide, is shown below.

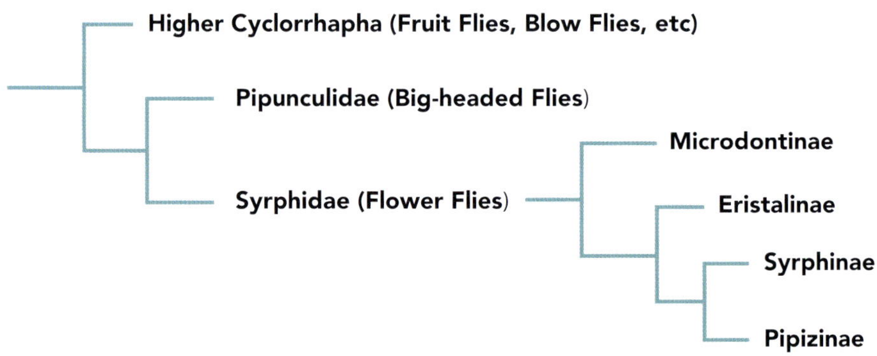

It may be of interest to point out that the most closely related of the other fly families, the sister family to the flower flies (Syrphidae), is the big-headed flies (Pipunculidae). These flies are very small, parasitic flies that have very large heads and are renowned for maneuvering through tight spaces while in flight, sharing a remarkable ability to hover. Other closely related families include the Lonchopterae, Platypezidae, and Phoridae, all part of the lower Cyclorrhapha taxon grouping of the Diptera.

FLOWER FLY BIOGEOGRAPHY

Biogeography of flower flies is another fascinating, and rapidly changing, field of knowledge. Very few clues to the location of the origin of flower flies exist, but there are indications that the first flower flies occurred either in Africa or the Orient. The story from origins to current distributions and diversity is very complex and only beginning to be understood. Part of the most recent phylogeographic story has to do with the strong ties between the Palearctic and the Nearctic faunas, indicating past dispersals, migrations, and colonizations across the various land bridges that have connected the continents over the last 55 million years. Both the closely-related species and Holarctic species point to ongoing and past connections between these to biogeographical areas.

FLOWER FLIES IN MINNESOTA

Minnesota's location at the center of the continent places it at the edge of a number of different biomes. This overlap and collision of different habitats account for much of the diversity of the state's flora and fauna. Minnesota is located on the western edge of the Great Lakes, the eastern edge of the Great Plains, and the southern edge of the boreal forests and the Canadian shield. These transitions are very noticeable when looking at a map of state's ecoregions.

MINNESOTA ECOREGIONS AND FLOWER FLY HABITATS

Level III ecoregions have been illustrated on the maps of this guide to give an idea of the diverse habitats of the state. At this level there are seven distinct regions:

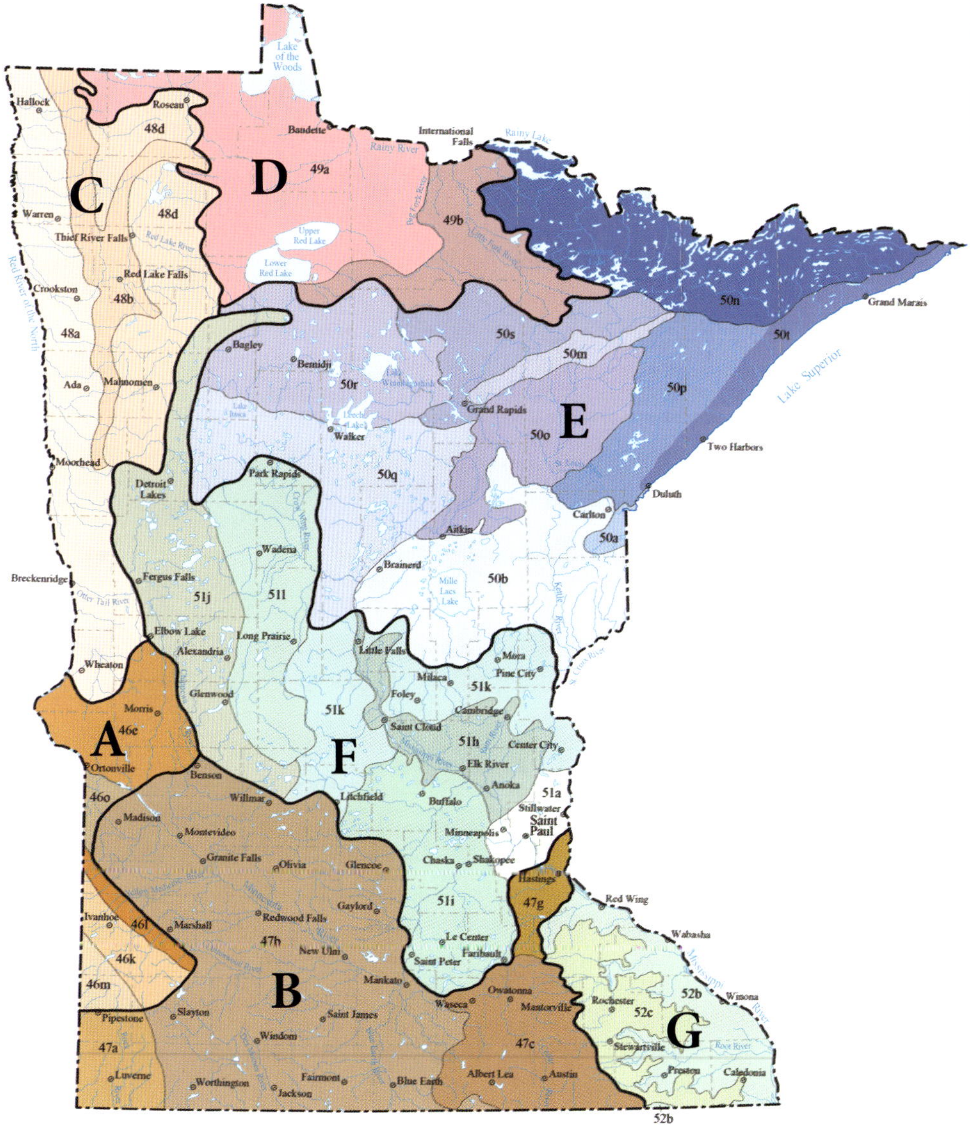

A. Northern Glaciated Plains: Prairie Coteau and Minnesota River Prairie.

B. Western Corn Belt Plains: Heavy agricultural lands.

C. Lake Agassiz Plain: Glacial Lake Bed and Beaches

D. Northern Minnesota Wetlands: Peatlands and Forested Lake Plains.

E. Northern Lakes and Forests: Boreal Forest and Boundary Lakes.

F. North Central Hardwoods: Deciduous forests and Moraine hills.

G. Driftless Area: Blufflands and Coulees. Trout streams.

FLOWER FLY FAUNA OF MINNESOTA

The present-day Syrphidae fauna of Minnesota is a result of large-scale climatic changes and subsequent faunal changes that have taken place since the last glaciation. This long-term history, spanning some 18,000 years, is complicated and difficult to piece together from the scant data. Much more is known about the acute alterations due to habitat disruptions, alterations, and decimations that have occurred over the last century and a half, and continue to occur. To monitor the changes resulting from these recent alterations to habitat it's crucial to have a good understanding of current and historical populations. Otherwise, there is no evidence of losses suffered or the effectiveness of conservation efforts. Unfortunately, the flower fly fauna of Minnesota is not well understood or well documented but our knowledge is improving.

Species that share similar distributions can be grouped into faunal elements. Most of the species found in Minnesota are restricted to the Nearctic. However, 18 species are **Holarctic** (9 Syrphinae; 1 Pipizinae; 8 Eristalinae), that is they have transcontinental distributions in both the **Nearctic** and the **Palaearctic** (Europe and Asia north of the Himalayas). And 6 species are adventive (all Eristalinae), having been introduced to North America from Europe.

The Common Drone Fly (*Eristalis tenax*) is a very large, Palearctic flower fly. This fly is adventive and was introduced to North America in the 1800s.

The Nearctic species of Minnesota can be further grouped into the following faunal groups:

1. **Northern** (Northern Boreal, Widespread Northern)
2. **Western**
3. **Eastern** (Eastern Transitional, Widespread Eastern)
4. **Southern**

Many species with eastern distributions reach their western or northwestern distribution limit in Minnesota. Likewise, many species with northern distributions reach their southern limit in Minnesota. Only a handful of southern and western species are part of the flower fly fauna of the state. The western species may well be underestimated because most of the collecting work has been in the eastern and northern parts of the state. And new southern species will no doubt arrive in Minnesota with future changes in climate.

A White-bowed Smoothwing flower fly (*Scaeva affinis*) feeding at flax flower. This is a western species, so far only observed in the northwest part of the state.

Syrphidae Faunal Groups by Subfamily

Subfamily	Northern	Eastern	Southern	Western	Undefined
Syrphinae	24	4	9	2	1
Pipizinae	3	3	0	0	0
Eristalinae	34	33	14	3	3
Microdontinae	1	3	2	0	0
Totals	**62**	**43**	**25**	**5**	**4**

FLOWER FLY MORPHOLOGY

The flower fly body is divided into three main parts: the head, the thorax, and the abdomen. Each of these parts has specialized appendages. The appendages of the head are the antennae and the mouthparts. The appendages of the thorax are the legs and wings. And the abdomen includes the genitalia.

Head

1. Eyes and Vision

The eyes are large, covering the sides of the heads. Males of most flower fly species have **holoptic eyes**, that is the eyes meet at the top of the head. Female eyes are **dichoptic**, a gap separates the eyes. This difference is often useful for differentiating the sexes. Even for the few genera where the males also have dichoptic eyes, such as *Helophilus* and *Sphegina*, the gap between the eyes is usually less than the gap between the female eyes.

A mating pair of Margined Calligraphers (Toxomerus marginatus) with the male (holoptic eyes) above and female (dichoptic eyes).

The large, compound eyes are a construct of numerous **ommatidia**, an array of tiny, individual lenses. Vision depends on the density and angle between these individual ommatidia. While usually red in color, often the eyes appear much darker, almost black in color, and less often they are light brown or yellow and patterned.

compound eye with numerous ommatidia

ocelli (simple eyes)

In addition to the two main eyes, like many other insects, flower flies also have three additional simple eyes known as **ocelli**. These are located at the top of the head between the larger main eyes and arranged in a triangle. These simple eyes are thought to monitor changes in overall light levels, changes that can trigger evasive flights from approaching predators, as well as roosting flights at dusk and initial flights at dawn.

2. Antennae

The shape of the antennae is often a first clue to recognizing flower flies in the field, separating them from bees and wasps which have long, many-segmented antennae, and other flies.

Flower fly antennae, while simple in structure, vary rather dramatically in shape and can prove quite useful in identification. The antennae are constructed of three main segments plus a hair-like **arista** located at the base of the third segment. The first segment is the **scape** and while it varies in length it is usually cylindrical. The second segment is the **pedicel** the end of which is concave. The third section is the **flagellum** (known formally as the funiculus), which attaches into the concavity of the second. The flagellum is most often lentil-shaped, and one of the distinct characters of the flower flies.

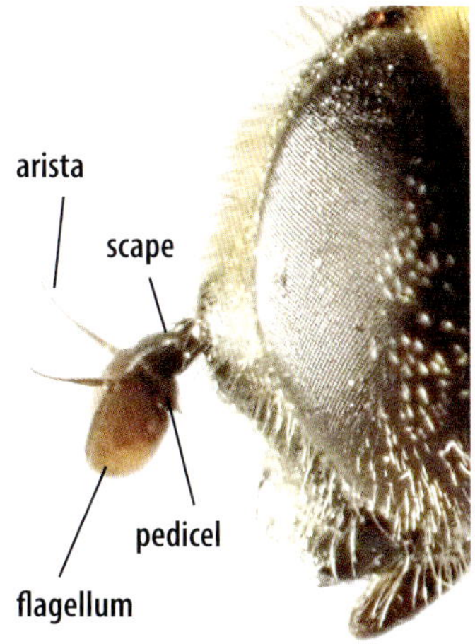

The arista arches above the flagellum to which it is attached. Most often it is attached near the base of the flagellum and dorsally but there are important exceptions. While almost always bare, several genera have plumose arista: *Sericomyia*, *Copestylum*, and *Volucella*. The shape and length of the various parts of the antennae are often used in identifying flies to genus and species. *See images next page showing some different antenna shapes.*

A Broad-banded Meadow Fly (*Chrysotoxum plumeum*), has an elongated scape, pedicel, and flagellum, causing the antennae to appear wasp-like.

A few genera have a prominence where the antennae attach to the head. The most-exaggerated prominence is that of *Sphiximorpha* which is nearly as long as the whole of the antenna. Several genera have an elongated scape, pedicel and flagellum, causing the antennae to appear more like the antennae of wasps, for example, *Chrysotoxum* and *Microdon*.

3. The Right Profile

The shape of the face, especially the profile, is also a useful guide to many of the genera. The flies of genus *Cheilosia*, for example, have a distinct knob on the center of their faces. *Volucella* have faces that are produced downward (*see photo below, bottom right*). Often a good photograph of a fly's face will aid in identification. The facial profiles for a number of common genera are shown below.

Brachyopa

Sphecomyia

Eristalis

Microdon

Spilomyia

Volucella

4. Mouthparts

The mouthparts of flower flies take on a variety of shapes and serve a variety of purposes. When not in use, the mouthparts are folded up. The **proboscis**, that part of the mouth that takes in liquids and solids, is complex and shaped differently for different genera. Flies such as *Rhingia nasica (right)* with extra-long mouthparts are specialists that can access flowers that have deep nectaries, flowers such as those in the mint family.

Thorax

The thorax is the largest part of the body. It is divided into three main parts—the **prothorax, mesothorax,** and **metathorax**—the exterior of which is built of a jigsaw-puzzle of **sclerites**, more than forty pieces have been identified and named.

One pair of legs is attached to each of the three main parts of the thorax. The front legs are sometimes referred to as **prolegs** and are attached to the prothorax. The middle legs are the **mesolegs** and are attached to the mesothorax. The rear legs are the **metalegs** attached to the metathorax.

Wings

1. Wings and Wing Venation

Wings are attached to the mesothorax above the mesolegs. Variations in wing venation often provide crucial information for identification. The **halteres**, golf-club like appendages and remnants of a second set of wings, are located just behind the wings. Two features of the wings are diagnostic of the flies of the family Syrphidae, features that are pretty much exclusive to flower flies. The first is the **vena spuria** or spurious vein. This is a structural, longitudinal fold in the wing membrane that visibly resembles the true veins. The second feature consists of a pair of marginal cross-veins which form a false-end to the wing margin.

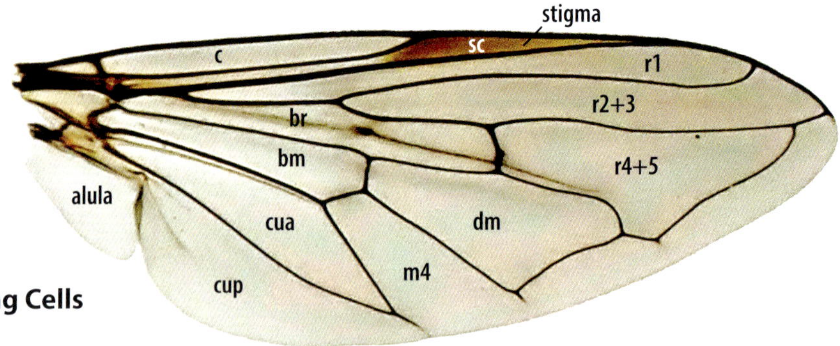

Wing Cells

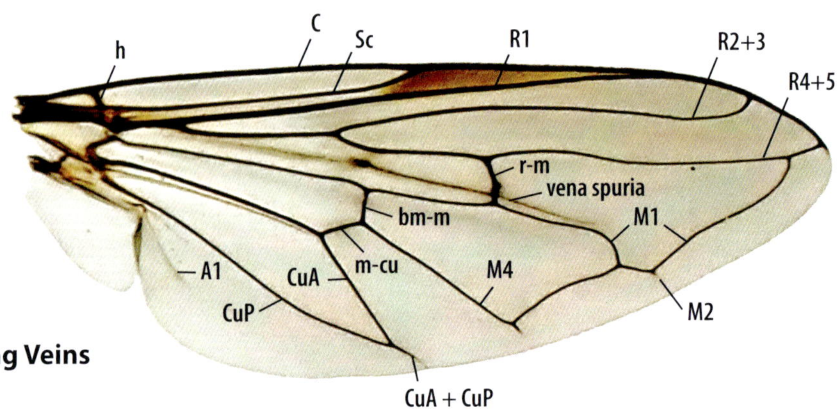

Wing Veins

Wing venation requires a rather complex terminology. The system used in this guide is indicated in the above figures. A number of wings from different genera have been photographed to be used as a reference and to give an idea of the differences (*see next page*).

The aerodynamics involved in the flight of flower flies is mind-boggling. The muscles that power the pair of wings and the wing structures work together to form a staggeringly complex system that is only just beginning to be understood, involving a combination of inclined stroke plane, wing twisting, camber and variable rotation. The gyroscopic role of the halteres is an interesting subject all on its own.

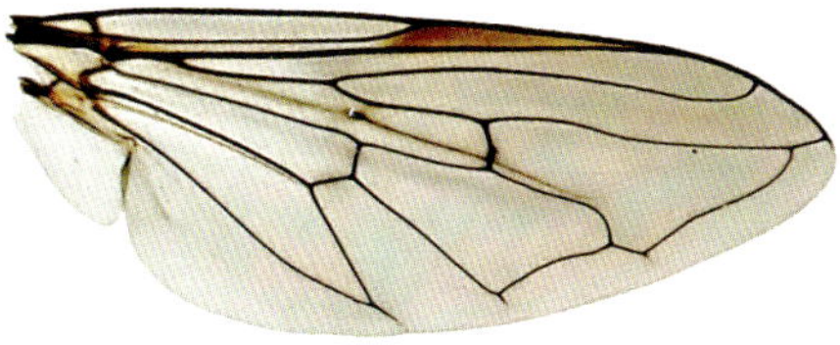

Chalcosyrphus nemorum

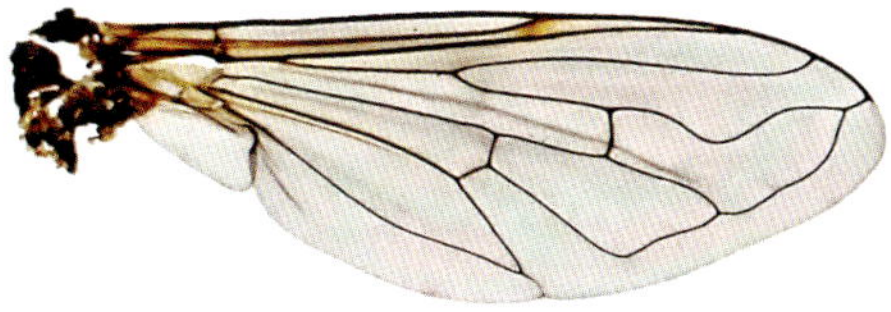

Eurimyia stipata

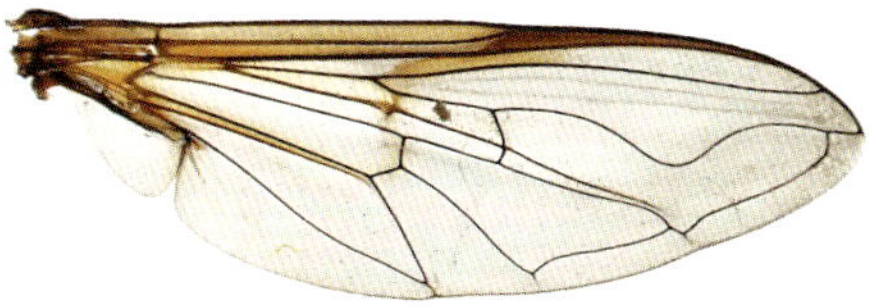

Didea fuscipes

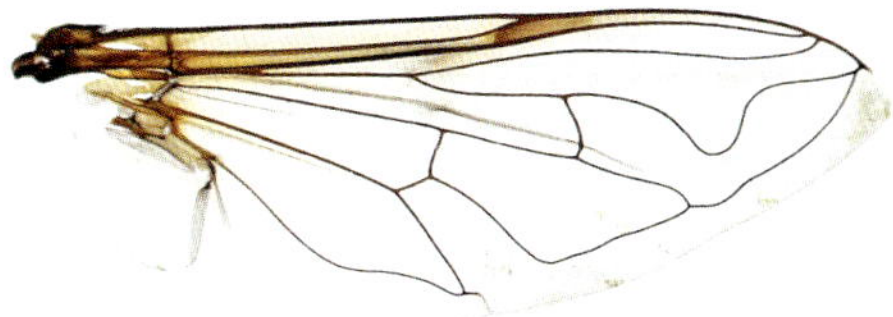

Helophilus fasciaatus

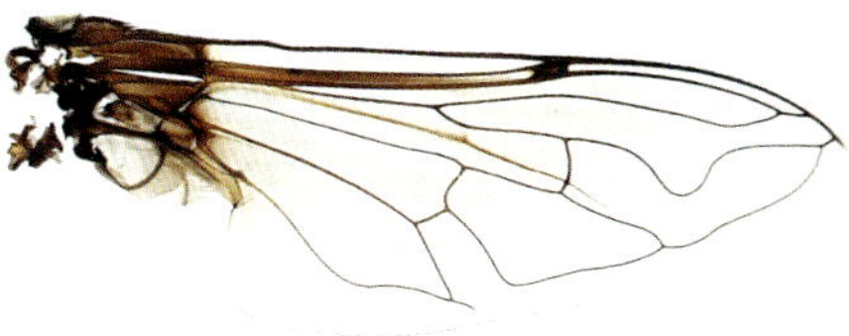

Eristalis dimidata

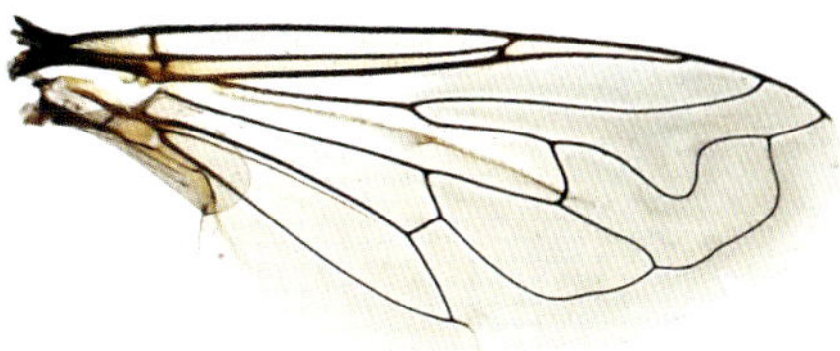

Merodon equestris

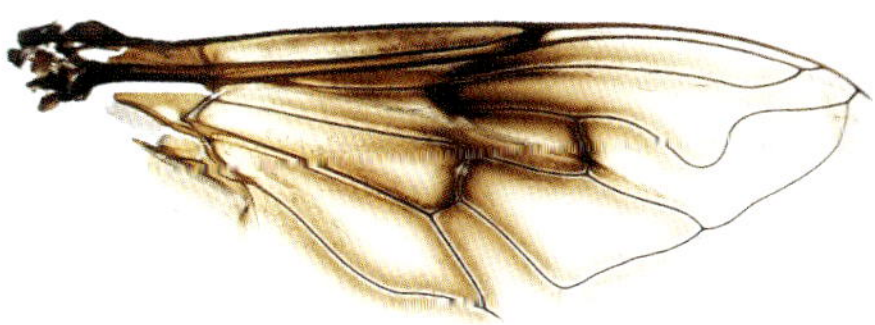

Eristalis flavipes

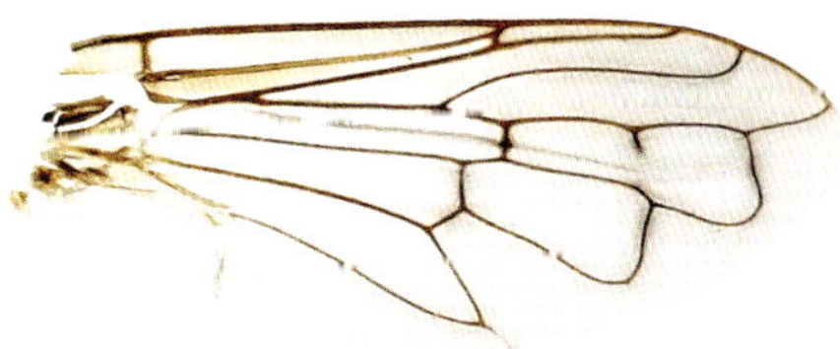

Microdon tristis

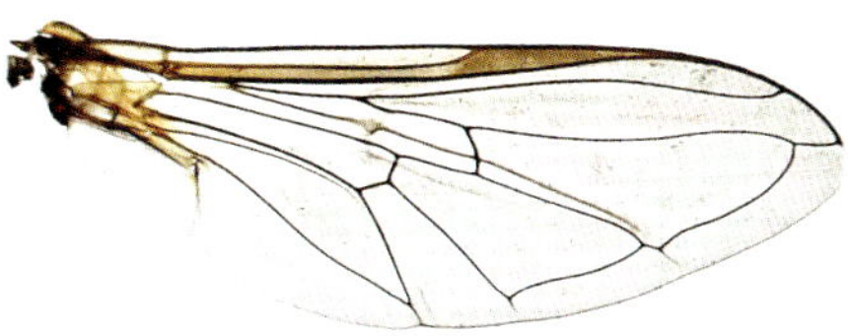

Eupeodes americanus

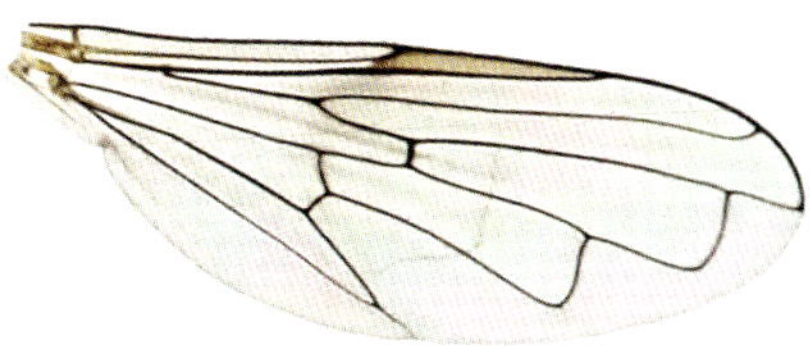

Neoascia sp.

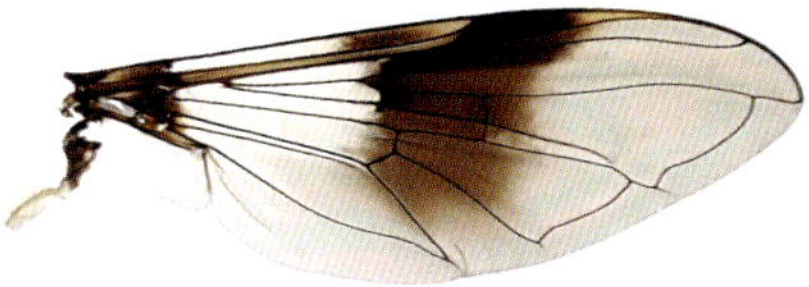

Hypocritanus fascipennis

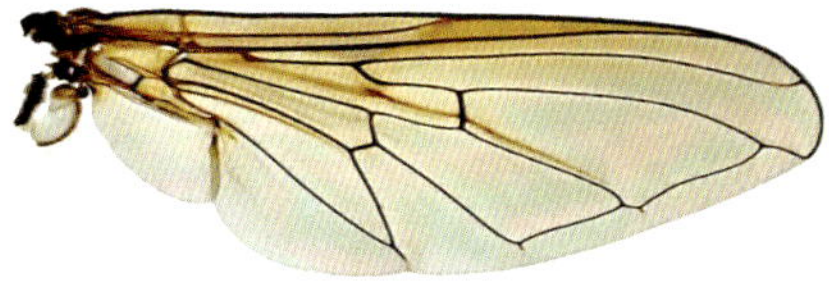

Rhingia nasica

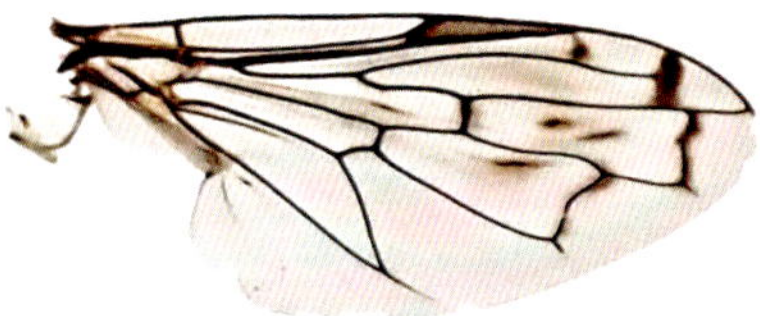

Orthonevra nitida

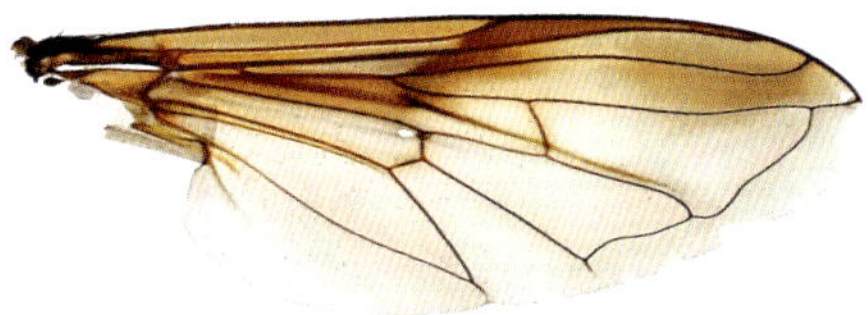

Sericomyia chrysotoxoides

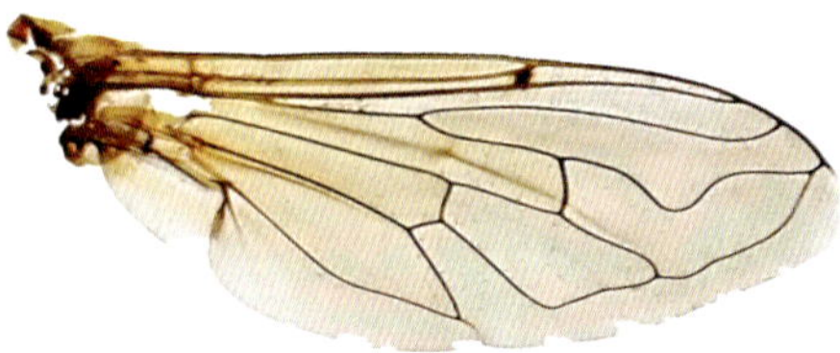

Parhelophilus laetus

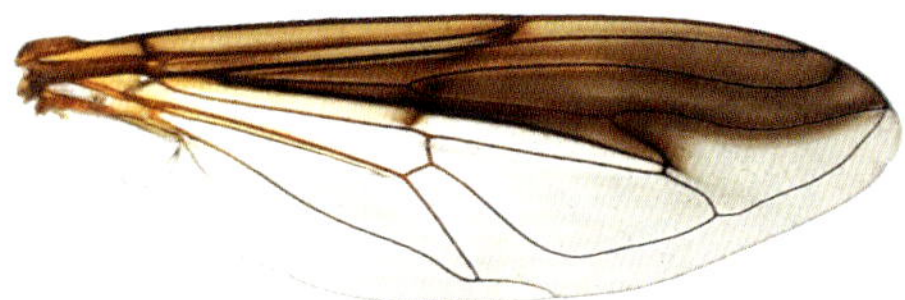

Spilomyia sayi

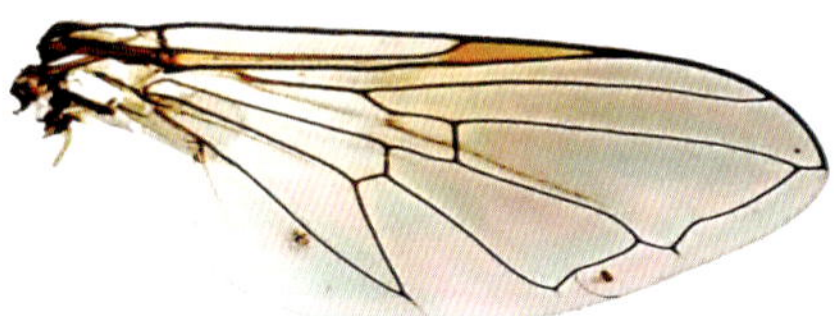

Pipiza femoralis

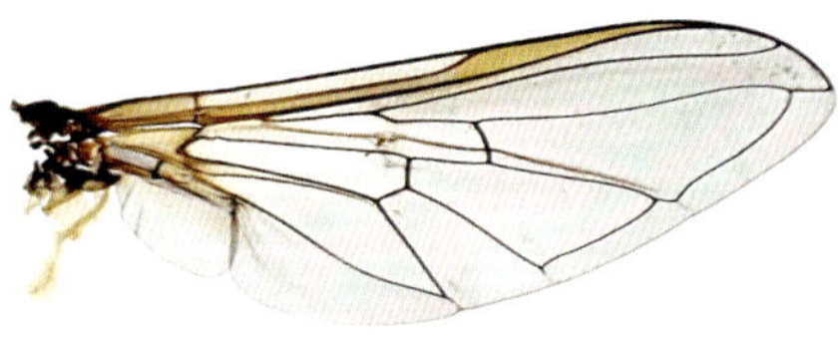

Syrphus ribesii

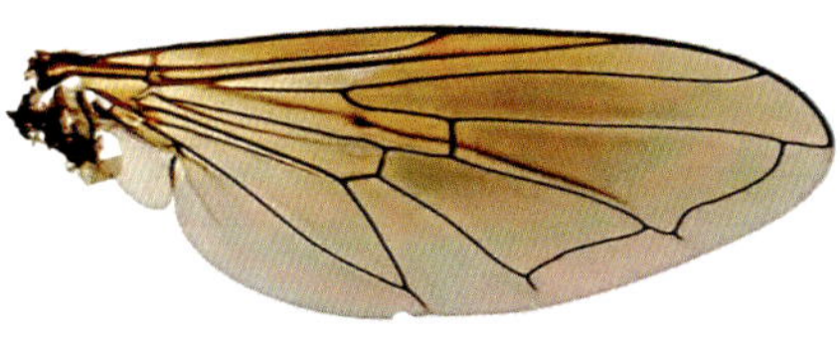

Platycheirus quadratus

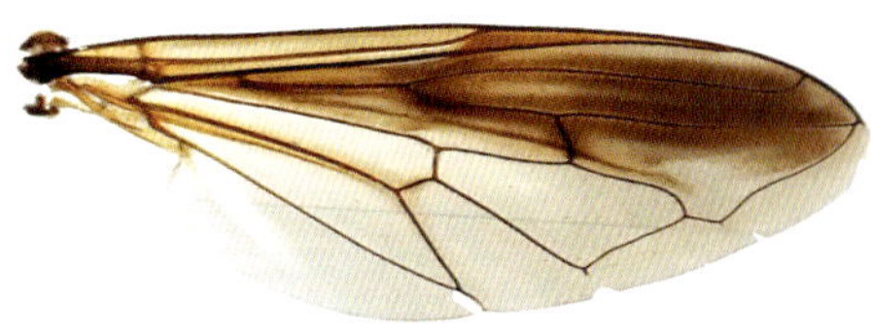

Temnostoma balyras

Legs

Flower fly legs are jointed and consist of nine named segments. Each leg attaches to the thorax beginning with two small segments: the **coxa** and the **trochanter**. The next segment is the **femur**, typically the largest part of the leg. Then the **tibia** and the five segments of the foot, known as the **tarsi**. These are numbered, with the 1st tarsal segment adjacent the tibia. The final tarsal segment, segment five, has a pair of claws.

The shape and coloration of the legs and leg segments are often useful for identification. The front legs of the Sedgesitters (genus *Platycheirus*) are curiously ornamented, with unusual hairs and flattened, strangely-shaped tarsi.

Abdomen

Flower fly abdomens consist of a series of wide segments. The 1st segment is mostly hidden beneath the **scutellum (metathorax).** The next three segments are the largest and make up the bulk of the abdomen. These segments often have distinct and colorful markings that are utilized for identification. The dorsal part of each section is a **tergite**, thus the segments are sometimes labeled T1, T2, T3 and T4. The plates on the underside of the segments are known as **sternites**.

The genitalia are located at the end of the abdomen. The males often have a visible capsule folded under the end of the abdomen. The female abdomens taper to a conical point that conceals the **ovipositor**.

MIMICRY

Mimicry is one of the wonderful byproducts of natural selection. Among the astonishing variety of insects, there is found an astonishing variety of mimics: from leaf butterflies and walking sticks that imitate the natural world, a kind of mimicry usually referred to as crypsis, to the unexpected similarities between unrelated species such as that between Viceroy and Monarch butterflies. The English naturalist Henry W. Bates described this form of mimicry in a paper in 1862 for butterflies he had studied in the Amazon. This type of mimicry, where an unrelated species acquires the warning colors and markings of a species well-defended from predators, is now known as **Batesian mimicry** in his honor.

Wasp-like Falsehorn (*Temnostoma alternans*)

Common Aerial Yellowjacket (*Dolichovespula arenaria*)

In the insect world, the order Hymenoptera which includes sawflies, ants, parasitic wasps, hunting wasps, bees, and social wasps are respected and avoided by most predators. This is due to the painful sting of the social wasps. In Minnesota, the pinnacle of pain rests with the bald-faced hornets and the several species of yellowjackets (all family Vespidae). These few species command so much respect that many other wasps and bees, which can also defend themselves by inflicting a sting to an attacker (though often far less painful), mimic the same bold colors. This type of mimicry is known as **Müllerian mimicry**.

Flower flies are well-known as excellent mimics of the Hymenoptera. Many are spot on mimics and difficult to tell apart at a glance in the field *(see photos above)*. Others are less convincing mimics, giving an overall impression of being bee-like. The mimicry of these latter species are being studied and have led to the science of "good enough" mimicry.

The adventive Drone Fly (*Eristalis tenax*) is a good mimic of the adventive Honey Bee (*see p. 145*). The Eastern Catkin Fly (*Brachypalpus oarus*) mimics the early mining bees (such as *Andrena* sp.) that share the same early maple and willow flowers.

There are a number of flower flies that mimic bumblebees. For example: *Eristalis flavipes, Mallota postica, Merodon equestris, Criorhina nigriventris,* and *Volucella evecta*.

The most impressive mimics are those that mimic the social wasps. There's a wonderful scene in Fredrik Sjöberg's memoir, *The Fly Trap*, where the author gets into a rather heated and comical argument about whether or not the *Temnostoma vespiforme* specimens he's collected are flies or wasps. The bicyclist on the other end of this argument remains unconvinced and rides away shaking his fist and yelling "Wasps!" Once you've encountered some of these mimics you'll understand just how convincing they can be. Some of our yellowjacket mimics include: *Chrysotoxum derivatum, Somula decora, Sphecomyia vittata,* and *Temnostoma alternans. Spilomyia fusca* and *Sericomyia militaris* are convincing mimics of the black-and-white bald-faced hornets. A number of other species, black with yellow markings, are good mimics of mason and potter wasps. These include: *Doros aequalis, Spilomyia sayi,* and *Temnostoma balyras.*

Orange-legged Drone Fly (*Eristalis flavipes*)

Red-belted Bumble Bee (*Bombus rufocinctus*)

Four-lined Hornet Fly (*Spilomyia sayi*)

Mason Wasp (*Ancistrocerus catskill*)

Another impressive mimic, the Violet Leafwalker (*Chalcosyrphus chalybeus*) not only looks like a spider wasp but behaves like one, moving restlessly about low vegetation, even flicking its wings.

Life Cycle Stages

Flies pass through several distinct stages as they complete their life cycle. The larvae grow through three stages, known as **instars**. The first, having just emerged from the egg, is the smallest. The third is the largest. When each instar outgrows its exoskeleton it molts. The third instar molts into a pupa. Inside the pupa, the larva transforms through metamorphosis into an adult fly.

Adult flower flies require the nutrients in pollen to become reproductively mature. Adult mosquitoes and deer flies require a blood meal to acquire the necessary nutrients. Adult robber flies get their protein another way, capturing and eating other large insects.

The Black-shouldered Drone Fly (*Eristalis dimidiata*) overwinters as an adult.

Most flower fly species complete their life cycle once in a given year. These are referred to as **univoltine** species. Some species, however, can complete two or more life cycles in a single year, that is they are **bivoltine** or **multivoltine**. Species that require two or more years to complete their life cycle, some of the *Microdon* species, for example, are considered to be **semivoltine**.

In Minnesota, resident species enter an inactive state known as **diapause** to avoid the harsh conditions of winter. Different species diapause at different life stages, some as larvae, some as pupae. A few species, such as *Eristalis dimidiata* and *Eupeodes americanus* are known to overwinter as adults.

1. Eggs

Flower fly eggs are small, elongate, and white or pale yellow. The eggs adhere to the substrate on which they were laid, and, typically, they hatch in a matter of days. The eggs of most species are not well documented or studied. Aphid eaters (genus *Eupeodes*) are known to oviposit single eggs on stems and leaves near aphid colonies, the future prey of the larvae. Most likely other **aphidophagous** species behave similarly. Marsh flies (genus *Helophilus*) are known to oviposit on leaves above standing water, into which newly-hatched larvae will drop and develop.

2. Larvae

The larvae of flower flies differ from the larvae of other flies in two distinct ways. The first unique character is the fused, posterior breathing tubes (in other flies the tubes remain separate). The second distinguishing character is the presence of longitudinal (rather than transverse) creases on the dorsal surface of the prothorax.

The larvae progress in growth through three stages, known as instars, with the third instar being the largest. The larvae differ in shape and structure depending on their preferred method of feeding. Larval feeding can be divided into three types: entomophagous, saprophagous, and phytophagous.

Entomophagous larvae feed on other insects. Species in the subfamily Syrphinae, in general, feed on aphids and other closely related hemipterans. These larvae are somewhat slug-shaped, being able to navigate and cling to the surfaces of plants where they feed. Often these larvae are colored and cryptically patterned.

The flies of the subfamily Pipizinae are similar, differing in that their larvae have specialized to feed on gall-forming aphids.

A slug-shaped larva in the subfamily Syrphinae feeds on aphids (entomophagous).

Species in the subfamily Microdontinae are also entomophagous but they feed on the eggs and larvae of ants and live inside ant nests. These larvae are so strangely shaped that they were not recognized as fly larvae but thought to be some kind of mollusk. They are flat on the bottom and dome-shaped above, often with an elaborate, geodesic texture. Probably this unusual morphology is connected to surviving in the hostile environment of an active ant nest.

Saprophagous larvae feed on microorganisms associated with decaying organic matter. This is the largest group and includes most of the flies in the subfamily Eristalinae (one notable exception is discussed below). Decaying organic matter occurs in a lot of flavors and accumulates in a number of places creating some interesting microhabitats. For example, sap runs on wounded trees, dead and decaying wood, rot holes in living trees, detritus in shallow water, even compost and manure. Different genera of the Eristalinae have specialized to take advantage of these microhabitats. Some of these situations result in very low oxygen levels and the larvae that feed there survive thanks to long, breathing tubes. These larvae are often referred to as rat-tailed maggots.

A rot hole in a living tree where decaying organic matter accumulates provides a microhabitat for saprophagous larvae.

Phytophagous larvae feed on the roots, stems, and leaves of plants. This is the least common method of feeding among the flower flies and includes flies from only three genera in Minnesota: *Merodon, Eumerus*, and *Cheilosia*. The first two genera feed in flower bulbs. The latter genus, *Cheilosia*, feed upon a number of plants and are often host-specific. Very little is known about the plant hosts and the fly-host relationships for the species found in Minnesota.

Research opportunity: Because so little is known about larvae of individual species, photographing and rearing of larvae will add to our knowledge of flower flies in Minnesota and elsewhere.

3. Pupae

The pupa (plural pupae) forms inside the **puparium**, the exoskeleton of the final instar of the larva. Once the larva pupates, metamorphosis begins. The pupal stage lasts days, usually a week or two, but may last months (if overwintering). Generally, larvae pupate near their preferred habitat—under bark, buried in soil, or well-concealed in leaf litter.

A *Chalcosyrphus nemorum* pupa.

4. Adults

When adult flower flies emerge from pupae they are quite vulnerable. Their exoskeletons are soft. And they cannot fly (see photos of *Chalcosyrphus nemorum*). Usually, the wings expand to full-size in less than an hour. The chitinous exoskeleton begins to harden during this time as well. Then, after the wings are expanded and dry, the adult flies off to a safer perch or to find food.

A recently emerged *Chalcosyrphus nemorum* with its wings not yet expanded.

Adult flower flies eat pollen and nectar. They require the protein and nutrients found in pollen to survive and reach sexual maturity. The adult flies utilize nectar as well—this food source is more of an energy drink, providing the needed pick-me-up for hovering flight, migration, and mating activities. The dependence on pollen and nectar requires the flies to visit many flowers and, as a result, makes them important pollinators.

Overwintering Strategies

Flower flies that overwinter as adults are usually migrants. *Eristalis tenax* overwinters as an adult and probably doesn't survive our coldest winters. Syrphines overwinter as third instar larvae in diapause. Little to nothing is known about the overwintering strategies of most individual species.

Diapause of flower flies has been little studied and is not well-understood. For insects that have been studied, once the required diapausal conditions are met, the remainder of the winter is spent in anticipation of warmer temperatures at which point development can proceed. An interesting (and probably observable) outcome of this process is that populations in the north will be on the wing as adults slightly later than populations of the same species in the south.

Helophilus fasciatus **perches on a leaf to bask in the sun in early spring. The adult flies of this species overwinter as adults thus are among the first flower flies to be observed in spring.**

Phytophagous species likely overwinter as pupae because their hosts die off and there is no food available to larvae in the spring. Saprophagous species overwinter in the larval stage, finishing their development in the spring.

FLOWER FLY BEHAVIOR

In general, adult flower flies begin their day in search of food. After perching overnight, a warm-up is necessary. They accomplish this by basking in direct sunlight or by wing-whirring, warming their body temperature by the vibration of wing muscles. When flight becomes possible, the flies visit their preferred food sources — flowers, sap runs, or honeydew on leaves. Later in the day, in addition to feeding, flower flies also perform other activities, also display other behaviors: males establish and defend territories; females search for breeding sites and oviposit; mating occurs (often in flight}; dispersal and migration take place, with some species congregate on hilltops.

Feeding

1. Adult Consumption of Pollen and Nectar

Adult flower flies eat the pollen and nectar of flowers. This is not just a preference as the nutrients contained in pollen are essential for the flies to become reproductively mature. The ant flies (subfamily Microdontinae) are a curious exception to this and apparently do not need to feed on pollen as adults.

Data for the association of specific flies with specific flowers are sparse and more observations will be needed before generalizations can begin to be made. Many species are generalists. While very few seem to limit their visits to a single flower species. The hairier the species probably the better they will be at transferring pollen grains. Examination of pollen grains stuck in the hairs of individual flies may be one way to make progress and add to our knowledge of which plants are visited by which flies. An interesting example can be gleaned from photos of *Helophilus* which show the large pollen grains of common milkweed stuck to their feet.

Research opportunity: Analysis of pollen grains on individual flies, either by microscope or DNA, would add to our knowledge of plant associations for different species. This would be especially helpful for species that feed in the forest canopy and that are rarely observed on flowers. This research could probably begin with specimens in insect collections.

Even though we know that almost all flower flies depend on flowers and the pollen and nectar they supply, very little is known about the foraging behavior of the flies. One exception is the Common Drone Fly (*Eristalis tenax*). This fly has several foraging patterns depending on the flower it's visiting. By documenting numerous visits, researchers discovered that the fly usually departs the flowerhead in the same direction from which it arrived. The flies usually circle a composite flower sampling mature florets and can be tricked into leaving the flower early or late by slowly spinning the flower. This suggests the fly uses visual cues to determine where it started and where it should stop.

Research opportunity: Study foraging behavior at flowers. Obviously this is not so easy as it sounds but at this point almost any observations will add to existing knowledge of flower fly behavior.

2. Larval Predation

Larvae of the subfamilies Syrphinae and Pipizinae prey on aphids. A few of the species and genera prey upon other soft-bodied insects as well, such as coccids and psyllids. The Pipizinae larvae are further specialized to feed on gall-forming aphids and psyllids. Adults lay eggs one at a time in or near active aphid colonies. The eggs hatch and the larvae begin feeding on aphids. The flower fly larvae stab their prey with sharp mouthparts then lift them up while eating them (*see photos of Eupeodes americanus*). The larvae pass through three instars, molting twice as they grow. During the third instar (after the second molt), syrphine larvae collect and store black, digested material in their hind guts. In some instances this black splotch of internal color aids in the larvae's cryptic coloration. Only when the third instar is done feeding and ready to pupate does it excrete this material, which leaves behind a tell-tale, tar-like deposit. While the predatory syrphine larvae may not be observed, the presence of these tar-like deposits on a plant is proof that they have been there feeding on aphids.

Larvae of the subfamily Microdontinae are also predacious but they feed upon ant larvae. These oddly-shaped larvae dwell inside ant colonies. When first discovered they were mistaken for some kind of mollusk. No doubt the hemispherical shape aids in protection from the ants guarding the ant colony.

3. Larval Saprophagy

Eristalinae larvae, with few exceptions, are saprophagous. These larvae feed upon the microbes responsible for the decay of dead organic matter, namely bacteria, yeast, and the hyphae of filamentous fungi. This activity facilitates the recycling of nutrients in the ecosystem.

Saprophagous larvae are found in a wide variety of habitats. Often a particular genus, subtribe, or tribe is associated with a particular larval habitat, displaying a close relationship to breeding sites. The dendrolimnetobionts, for example, live only in rot holes in mature trees. This group includes flower flies from the following genera: *Myolepta*, *Spilomyia*, and *Mallota*. *Criorhinia* require hardwood forests with ample decaying leaves. *Temnostoma* are associated with decaying wood. The larvae of *Copestylum* feed in decaying cactus. *Volucella* larvae are scavengers inside large, social wasp nests, living and feeding among the debris and waste generated by the wasp colonies.

Research opportunity: rot hole monitoring and emergence traps.

Hovering

Flower flies are renowned for their flying ability and in particular their hovering. Motionless motion. Their wings moving so fast as to be invisible, the flies remain in place in mid-air for long periods of time. Often this is the male fly defending some choice location. Some flies return day after day to hover in the exact same location. Syrphines move about, hovering in one location, darting off, then hovering again in a different location. Some hover high above the ground. Others remain close to the ground or move like magic through caves and passageways in dense vegetation. It's an impressive feat and one worth stopping to admire when encountered.

Hilltopping

Many flower flies, especially uncommon species, congregate on hilltops and other prominent spots. This behavior is thought to increase the likelihood of finding a mate for species that have low numbers of individuals. These locations can be excellent places to look for and observe flower flies, especially rare flower flies. Search these sites early in the day.

Migration

As with most insects, there exist a range of movements. Daily, localized movements from roosting sites to feeding sites to breeding locations. Longer flights from areas of high population density to lower population density, colonizing new habitats, establishing satellite populations. Accidental movements due to severe weather. And long-distant, seasonal, intentional movements considered to be migrations. Very little is known or documented about the movements of Minnesota flower flies. Many European species are known migrants. It would be unrealistic to think that none of our species share this same trait.

THREATS TO FLOWER FLIES

Flower flies have numerous natural enemies and have evolved a number of ways to prevent being preyed upon by birds, amphibians, reptiles, and other arthropods such as spiders, dragonflies and wasps.

Birds prey upon both adult flies and the larvae. Dragonflies prey upon adult flies. Adult flower flies, small syrphines in particular, are preyed upon by the hunting wasps of the genus *Ectemnius*. These small wasps adeptly capture flies while they are hovering. The large sand wasps of the genus *Bembix* hunt larger flies. These impressive wasps capture and provision their nests with horse flies and deer flies but also capture many flower flies, especially drone flies (genus *Eristalis*).

The main method of escaping and avoiding predators is by flight. The Syrphini, flies such as *Eupeodes* and *Syrphus*, for example, are very strong fliers. When threatened they disappear high up into the canopy of nearby trees. These flies are also rather skittish and if they don't just leave the scene, they move away from the sunlit edge of vegetation into the shadowy undergrowth, maneuvering easily among the thick stems and leaves.

Other important methods of avoiding predation among the flower flies are mimicry, crypsis, and camouflage. Many flower flies are colored with bright yellow markings and are convincing mimics of wasps while many others are bumblebee and mining bee look-alikes. Predators that have had painful encounters with the authentic, stinging originals tend to leave the mimics alone. Other flies are colored to blend into their surroundings.

Flower fly larvae are targeted by a variety of parasitic wasps. The Ichneumonid wasps of the subfamily Diplazontinae attack the larvae of the aphid-eating larvae of the Syrphinae. In Minnesota, there are records of *Diplazon laetatorius*, which has one of the greatest geographical ranges of any wasp. This is most likely due to it being spread by agriculture along with its host flower flies and the flower flies' host aphids. Flower fly larvae that live in rot holes are often targeted by the predatory larvae of the muscid fly genus *Phaonia*.

***Sericomyia militaris* captured by a flower crab spider.**

A jagged ambush bug captures a *Toxomerus* sp.

A square-headed wasp (*Ectemnius maculosus*) stings her prey, a soldier fly, captured on a flower.

A sand wasp (*Bembix americana*) with her captured flower fly prey clutched beneath her.

Above and beyond the whole host of natural threats to flower flies, the greatest threat to flower fly populations is habitat loss. So many current land-use practices and land use management policies are simply incompatible with the needs of flower flies in general. Agricultural, domestic, and industrial interests rarely account for much beyond human interests. One only needs to reflect upon the vocal resistance to buffer zones in Minnesota and the somnambulistic use of yard chemicals to realize the scope of the threat against native flora and fauna. Add to these established negative pressures climate change and one begins to feel the corners and walls that native plant and animal populations are driven up against.

The wild is everything that grows and breeds and functions without supervision or imposed control. It is what lives in the long term without being managed.

Robert Bringhurst, *Learning to Die: Wisdom in the Age of Climate Crisis*

So ingrained is the concept of management that in Britain we do not seem very interested in how the natural world actually works.

Peter Marren, *Nature Conservation*

CONSERVATION AND CLIMATE CHANGE

Here's the thing: Native plants and their pollinators have lost much of their previously-available habitat. Pre-settlement, pre-industrial populations existed across large tracts of land with abundant, connected, relatively-stable habitats. These landscapes have been altered, and continue to be altered, with suitable habitats reduced almost out of existence in some instances. The populations of most plants and animals have been displaced, pushed aside, and relegated to marginal habitat and remnants of land, small areas overlooked or off-limits. In the southern, agricultural part of Minnesota and the large urban centers, habitat loss or alteration approaches 100%. There are very few acres of untouched or unmanaged land. There's no way change of such sweeping extent doesn't affect the ecology of these regions. Even if good records don't exist of what has been lost, losses in number and diversity have surely occurred.

In addition to drastic reductions in suitable habitat, insects also suffer from ill-founded attitudes, attitudes perpetuated by cultural traditions and marketed by pest control businesses. Insects, and pretty much all insects that aren't butterflies or dragonflies, are regarded at best as nuisances and at worst as pests that need to be eradicated, this despite the fact that the vast majority of insects are beneficial, and not just beneficial but essential. Such attitudes make conservation difficult. This popular animosity toward insects undermines and opposes many of the goals of conservation.

What Remains

Somehow, despite devastating habitat loss and the continuous threat of chemical extermination, all is not lost, nor is it too late to assist remaining pollinator populations. A truly remarkable number of species persist, though no doubt in diminished numbers and more localized populations. Many parks and nature reserves now exist, reclaiming previously farmed or forested lands. In addition to federal, state, and local lands set aside for public use and for conservational reasons—many weedy, ungroomed, untended margins harbor a disproportionate share of our flora and fauna—road ditches, property boundaries, shorelines and the like.

A reconstructed prairie can provide critical habitat for flower flies and other essential pollinators.

Just as a small percentage of people control a majority of the wealth, a small percentage of undisturbed and unmanaged lands harbor an improportionate share of the diversity of plants and animals. Of course, there's a dark irony in this comparison, given that the economy generating this wealth is linked directly to habitat loss due to its often unchecked exploitation of natural resources. With the economy and the natural world at cross-purposes, the importance of protecting remaining habitats cannot be overemphasized. Nor can the importance of adding areas of new habitat to the little that remains be understated.

Fortunately, the very fact that small refuges harbor a wide variety of plants and animals suggests that additions (no matter how small) to existing habitat can be beneficial. So, while the wide-angle view may be bleak and discouraging, it's important to realize small actions can be taken that will aid flower flies and other pollinators.

What Is To Be Done?

So, what can be done to help out our flower flies, these interesting and beneficial insects that, for the vast majority of people, barely exist? In fact, there's something paradox-like about this question, a koan-like riddle with no solution. It has to be doubted whether the plight of any obscure species of fly will stir the population or state to any level of economic sacrifice. The answer requires breaking-through doubt to some measure of understanding and action, some balance of regrets and appreciation. The answer requires a kind of awakening.

One of the aims of this guidebook is to introduce these flies and raise awareness of their existence and of their beneficial roles. Knowledge of an important (but previously unknown) piece of our world can often be the first step toward conservation. Aldo Leopold hoped, in the 'Land Ethic' chapter of *A Sand Country Almanac*, that property owners, if they became more ecologically-minded (you can read this as flower-fly-minded in our case) they would be proud to act as stewards of a reasonable portion of their holdings in support of diversity and beauty, for the betterment of their property and the community. This is the ethical path to conservation.

Of course, conservation of any species turns not only on the willingness of the property owner but also on the preservation of suitable habitat. While protective legislation specifically aimed at flower flies is not realistic until a better understanding of the populations and distributions of the flower flies in Minnesota is established, conservation measures taken to protect and maintain habitats for "higher" species—mammals, amphibians, birds, and rare plants—shouldn't harm the populations of flower flies. Unknown causes can piggyback upon known causes, the ad hoc path to conservation.

1. Pollinator-friendly plantings.

Rain gardens, pocket gardens, prairie plantings, road ditches, fence lines, wood lots, and field margins can provide important habitat to flower flies (and other insects), Existing habitat can be created or improved by replacing invasive monocultures of grass or non-native plants with a diversity of native plantings.

A church rain garden provides habitat for flower flies and pollinators, and functions to capture water runoff from the parking lot, preventing pollutants from getting into nearby rivers and lakes.

While the propagation of native species of plants and trees are particularly helpful. A couple additional guidelines should be considered: Planting large clumps of a single species will be better than single plants of differing species. Choosing a variety of plants with a variety of flower shapes (i.e. from different plant families) will attract insects with different mouth shapes and feeding habits. Providing plants that bloom at different times of the year, ideally with something blooming throughout the seasons without any gaps, will ensure food sources for a succession of pollinators.

Trees provide essential habitat for many species of flower flies. Unlike lawns and manicured green spaces, the upper limbs and leaves are seldom subjected to the barrage of chemicals applied routinely at ground level (often at a maddeningly-higher rate than agricultural lands!?). As such, the tree canopy remains—even in urban settings—one of our true wild habitats, mainly because it remains out of reach of our routine "improvements."

2. Land Management Practices

There are several specific habitats that are somewhat unique to flower flies that should be taken into consideration when making land management decisions. In most instances, these qualify as micro-habitats, nested within larger habitats. Additionally, the majority of these micro-habitats are associated with mature forests and high-quality wetlands. The small, aquatic micro-habitats that exist inside rot-holes on living trees. Sap runs and leaks on living trees. Decaying sap and phloem layers under the bark of freshly fallen trees and tree limbs.

> **So I imagined a forest world. A world of plants, interconnected by root systems, pollen drift, and other interchanges and more ethereal linkages so complex as to rise to full sentience and perhaps intelligence.**
>
> Ursula Le Guin

Probably too many trees in cities and on private and managed property are cut down prematurely, essentially depriving many native insects, including flower flies, their specialized habitats. The insects that develop in these aging trees feed numerous species of birds as well. Not only do the flies visit and pollinate tree flowers but the larvae of many species depend upon trees as well, many feeding upon tree aphids or living in rot holes. The way forest wood lots, city parks, and urban trees on private property and along city streets are tended will almost certainly impact populations of many flower flies.

Buffer strips, cover crops and other land-use strategies aimed at conserving topsoil will also benefit native pollinators: the flowering plants providing a food source and the undisturbed ground providing nesting habitat. Xerces Society and the Land Stewardship Project are two organizations that can provide information about best practices for landowners.

Interestingly, the possibility exists of creating artificial flower fly microhabitats. This has been done in Scotland as a way of surveying for certain rare species. Holes are cut into the top of a tree stump or block of wood and left to fill with rain. The resultant small pool provides a suitable breeding site for a few species. A substitute rot-hole can be created from a plastic bottle with a round hole cut in the side and the bottom filled with wood chips, bark, twigs, leaves, and water. The bottle suspended in a tree.

3. Learn More About the Complex Workings of the Natural World

Books such as *Bees, Wasps, and Ants: The Indispensable Role of Hymenoptera in Gardens* by Eric Grissell (he also wrote the wonderful, and just as relevant, *Insects and Gardens*), *No Nettles Required: The Reassuring Truth about Wildlife Gardening* by Ken Thompson, or *Plant Galls* by Margaret Redfern can provide mind-altering information about many overlooked and underappreciated parts of the natural world. There are many other such books. Online bookstores, such as Pemberley Natural History Books, are great places to browse and follow your curiosity.

The iNaturalist phone app (when used properly!—the potential for misuse or mistakes are great) uses machine learning to suggest possible identification of photographed plants and animals. This provides a welcome (and lightning-fast) compliment to guidebooks. The resulting Artificial Intelligence, it must be remembered, makes mistakes and isn't always accurate. The fact that it works as well as it does results from well-vetted observations, good photos confirmed by knowledgeable experts.

Keep a journal.

In addition to taking notes of the weather and the time and location, all useful information, a field notebook is a great resource, allowing one to write down questions and stray thoughts with the benefit of being able to revisit these jottings later. Detailed notes about behavior can be extremely valuable, as this information is rarely captured or conveyed in a photograph or with a collected specimen. The physical act of writing down names and observations will also aid in learning the names and building pattern recognition.

Climate Change

Changes to Minnesota's climate will impact plants and animals in numerous ways. The climate in Minnesota is changing, rapidly warming and receiving more rain. According to the Minnesota DNR the top ten years on record for average temperature and total precipitation have all occurred since 2000. Winters are warming more than summers, allowing the persistence of plant and animal species that, previously, could not survive the harsh winters. In addition to an overall increase of precipitation, we are also experiencing larger, more-prolonged events—for example, again since 2000, rainstorms delivering six inches of rain or more over wide areas have increased fourfold. Slight changes in scale, duration, and intensity of weather events lead to large changes on the ground. More and more frequently it doesn't rain but it pours.

Another concern is the change in growing season or frost-free season. The currently increasing growing season affects the yearly life-cycle of both plants and animals as well as the intricate and complex relationships between them. For some plants, the timing of their blooms and the presence of their pollinators could be disrupted resulting in a reduction of fruit and of fertile seeds needed for propagation. The timing of aphids and their natural predators, the aphideaters, could be affected to the detriment of the plant hosts. These are but a few of the uncountable consequences of climate change.

Spring-flowering plants such as Virginia spring beauty (*Claytonia virginica*) that provide pollen for flower flies may be affected by climate change, flowering earlier or later than historical records.

Yet another pressure on flower fly populations in addition to those of habitat loss and pesticides. In England, a number of flower fly species have expanded their range northward in the last two decades. What climate change may mean to Minnesota flower fly populations remains uncertain. Without the observation data that exists in well-surveyed countries like the U.K., we can only speculate that southern species will expand northward into Minnesota and the boreal species may disappear from northern parts of the state as the range for those species is pushed farther to the north.

STUDYING FLOWER FLIES

There's something unpretentious about the study of flower flies. Just try to explain to a passerby that you are photographing or, worse, watching flies. While many of the flies are indeed flashy enough to be considered beautiful by most standards, they're still flies, and as my wife pointed out to me, most people own fly swatters.

1. Approaches to Studying Flower Flies

The Japanese haiku master, Basho, instructed: "Learn about pines from the pine, and about bamboo from the bamboo." Certainly, the surprise of the encounter is lost. As are subtle points of context and behavior, knowledge of which accumulates over multiple encounters. In the end, you need to see these flies for yourself. The best way to learn about flower flies is from the flower flies.

Observing Flower Flies

Flower flies, like the bees and wasps they often mimic, move about almost constantly. This incessant movement, from flower to flower, from leaf to leaf, can make them a challenge to observe and identify. In addition to flitting about, most flower flies are very small. With practice, many species will be recognizable by sight alone. The assistance of a camera, close-focusing binoculars, or a net will be useful for many of the flies.

Photographing Flower Flies

While photographing flies is not easy it is rewarding. Many of the species are quite showy and wonderfully colored, especially the large wasp mimics. In addition, digital photos can be uploaded to citizen science platforms for identification and to contribute to known distributions of species. The fact that these photos can be enlarged makes them useful for observing characters that are difficult to see with just eyes alone.

Flower flies move frequently. And if there is any breeze at all the flower they are visiting swings back-and-forth like a metronome. Most of the photos in this guidebook were taken using an Olympus micro 4/3 camera with a macro lens. I routinely use the burst setting. The rapid-fire photos results in a succession of shots with, hopefully, one or more in focus. One of the luxuries of digital photography is that one can overshoot, taking hundreds of photos of a single subject. One of the labors of digital photography follows, sifting through all the photos and deciding which of the multitude to keep.

Collecting and Preserving Specimens

Collecting flower flies remains an important part of the study of these insects. Preserved voucher specimens provide the material used by taxonomists, phylogeneticists, and other biologists to construct our scientific knowledge of the flower flies. Without their work and the preserved specimens, guidebooks, such as this one, would not be possible.

A hand net, vials, kill bottle, pins and tightly sealed storage drawers are the basic equipment needed to begin collecting. All these necessities can be purchased from online entomological suppliers such as BioQuip, NHBS, or Acorn Naturalists. Netting a fly is often advantageous, allowing close inspection of the living fly with a hand-lens or posed macro photographs. The fly can be released after examination or killed and preserved as part of a reference collection.

Flies can be killed using ethyl acetate in a kill jar or frozen. The flies must be pinned, dried, and labeled as soon as possible. Labels must include, at a minimum, the following information: date of collection, location (GPS coordinates, county, state, nearest town, etc.), and collector's name. Without this information, the specimen is not useful. Additional information (often added as additional labels) includes species name, sex, habitat, plant associate, behavior, or collection number.

Once a collection is begun, measures must be taken to preserve the specimens and make use of the information. Put another way, if a fly had to be killed for a voucher specimen steps should be taken to make it as useful as possible. The standard storage unit, the Cornell University drawer, provides the best protection, solidly constructed of wood with a tight-fitting glass top. Cardboard unit trays with foam bottoms for pinning are arranged inside the drawer, making it easy to organize a collection by genus, by species, or by some other taxonomic designation. Some insecticide should be kept inside the drawers—mothballs or pieces of no-pest strips. Alternately, the drawers (and contents) can be placed in extreme cold and routinely frozen. This is a convenient approach in Minnesota where leaving the collection outside for a day or two during a cold snap in January should kill most pests.

Arrangements should be made to donate the collection to a larger, permanent collection such as the University of Minnesota Insect Collection or the Science Museum of Minnesota. This will ensure that the specimens are available for study in the future.

2. Where to Look

Edges

Flowering plants and shrubs at the edge of openings and fields adjacent to mature forests are very good places to begin looking for flower flies.

Forests

Species associated with the forest canopy can often be observed closer to the ground in sunlit glades—basking on leaves, feeding on honeydew, searching for mates, visiting flowers, and occasionally landing on bare ground for water and minerals. Species such as *Brachyopa* and *Hammerschmidtia* will most often be found near their preferred larval habitats—sap runs on old trees and fallen aspen logs, respectively. The ant flies, a group of flower flies that do not visit flowers, should be sought near the ant nests that the larvae live in.

A Black-bodied Ant Fly (*Microdon megalogaster*) basks on a leaf in a forest.

Public Land

Minnesotans are fortunate to have access to a great abundance and variety of public land, all of which provide some suitable habitat for flower flies. These are great places to look for flower flies and very few have been surveyed. Be advised that many of these locations require permits for scientific collection of insects. Thankfully no paperwork is required to take photographs.

These public lands include:

- Wildlife Management Areas
- State Parks (more than a quarter-million acres in 75 parks and recreation areas)
- Scientific and Natural Areas
- The Boundary Waters Canoe Area Wilderness
- National Parks and Forests
- National Wildlife Refuges
- NGOs like The Nature Conservancy, National Audobon Society, Ducks Unlimited and Pheasants Forever
- County Parks
- City Parks, and
- local nature centers.

Most city parks tend toward the overgroomed—ball diamonds, playgrounds, mown green-space—and these are the least supportive of any wild, native, or natural diversity. This is—literally—not markedly different than a parking lot at a mall or hundreds of acres of corn. These "spruced-up" areas should be the very last place to look for flower flies. Even still, most of these parks often overlook some areas and have frayed edges—untended margins—maybe a riverbank or railroad grade, or an adjacent lot of messy, dilapidated, second- or third-growth woods. These unkempt edges and scruffy margins are usually worth a visit.

Gardens

More and more, the city parks have set aside areas for gardens. Flowers in these gardens will certainly attract a variety of flower flies. As will plantings on your own property. Our small city plot currently has about sixty species of flowering plants in small gardens around our house. To date, thirty-four species of flower flies have made an appearance on these plants. It's a real pleasure to be able to step out the door and observe flower flies in your own yard.

3. When to Look

Phenology

A graph of all flower fly observations by date on iNaturalist yields a curve with two distinct peaks, one in June and another in September (*see graphic below*). This indicates that there are at least two groupings of species: early spring species and late summer species.

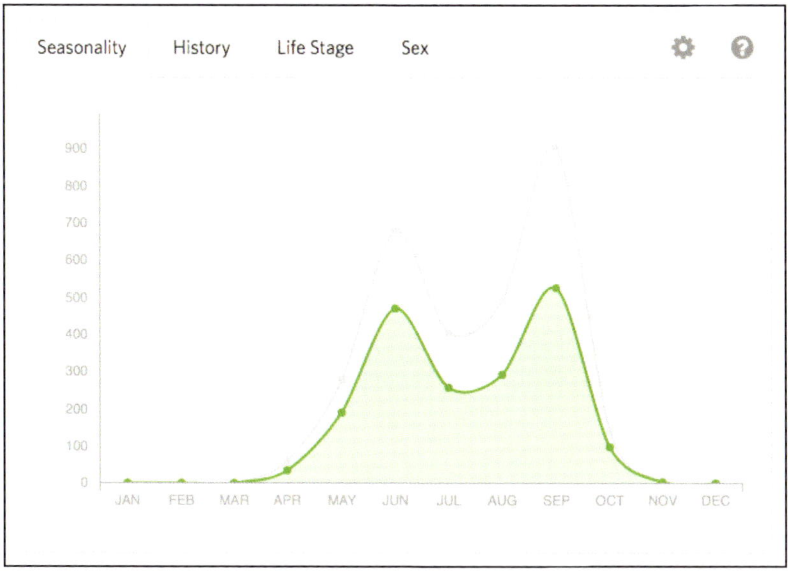

An iNaturalist graphic showing the bimodal seasonality of the syrphid flies for North America.

Even so, once the first flowers bloom in the spring, the flower flies will be on the wing and will be until the snow flies in autumn. The succession of blossoming plants will attract a succession of pollinators; bees, beetles, and flies. Following the bloom from spring to fall—spring ephemerals, willows and maples; raspberry, dogwood, and sumac; thistle, goldenrod and aster—will yield a variety of flower flies.

Time of Day

The time of day matters as does the weather. Many flower flies, like other insects, don't begin to fly until the temperature reaches 60 degrees F. Sunny, late mornings seem to be the optimal time for observing flower flies.

4. Flower Fly Study in Minnesota

Many fascinating people have contributed to our knowledge of the flower flies of North America. Near the beginning was Carl-Robert Romanovich Baron von der Osten-Sacken. Osten-Sacken was not only a polyglot Russian diplomat with a fantastic name but he also kept a life-long passion for flies. During his years in the United States (1856 - 1877), Osten-Sacken tirelessly collected fly specimens. While his main interest was crane flies (Diptera: Tipulidae), his catholic collecting habits and his long collaboration with the German taxonomist Hermann Loew (to whom he sent many of the flies he collected) resulted in the first descriptions of many of Minnesota's flower flies (something close to one-fifth of the species on the current state list bear the name of Loew or Osten-Sacken). Subsequent to the publication *Diptera of North America* in 1877, Osten-Sacken's massive collection of flies was shipped back across the Atlantic by Loew and returned to the Museum of Comparative Zoology at Harvard in Cambridge.

Samuel Wendell Williston began his career as a field collector for paleontologist Othniel Charles Marsh at Yale University. This was during the time of Marsh's infamous, wild-west-rivalry with Edwin Drinker Cope, now known as the bone wars. Williston chafed under the tyranny of Marsh, and instead of quitting (which he threatened to do many times), Williston branched out into an area of science not under Marsh's dominion, namely the study of flies.

So, during the escalation of the Bone Wars, Williston diversified, beginning his study of the family Syrphidae. Most impressively, Williston acquired his knowledge of flower flies almost entirely on his own, through persistence and sheer willpower. These efforts culminated in the publication of *Synopsis of the North American Syrphidae* (1886). While this work includes very little content about Williston's days in the field or about the natural history of the flies, it does include the following passage "They are flower-flies, and feed upon honey and pollen. They are observed on blossoms of sweet-smelling, melliferous plants, such as the Hymenoptera prefer; and patches in bloom of blackberry (*Rubus*), wild cherry (Prunus), dogwood (Cornus), Canada thistle (*Cirsum*), and elderberry will always be sure to reward the patience of the collector."

From 1933 to 1941, Horace S. Telford was a graduate student in entomology at the University of Minnesota. His research culminated in the publication *The Syrphidae of Minnesota*. After graduating, Telford worked as an entomologist in North Dakota, Michigan, and Ohio before finally settling in Washington, where he served as professor of entomology for nearly thirty years at Washington State University in Pullman. After his retirement in 1975, he conducted research in Brazil and continued to teach at Gainsville and Florida A&M Universities. Publication of a checklist of the flower flies for the state of Washington in 1975 is proof that Telford kept a life-long interest in this group of insects. It's interesting to add that Telford continues to be remembered for conducting some of the first studies, in the 1940s, that confirmed the dangers of DDT.

Four decades after Telford's publication, John Haarstad began his work as a naturalist at the Cedar Creek Ecosystem Science Reserve in east-central Minnesota. Over several decades, Haarstad would amass an unparalleled reference collection of insects, especially rich were his collections of Odonata, Hymenoptera, and Diptera, including flower flies. To Haarstad's lasting credit, he photographed many of the pinned specimens and published the material to the Cedar Creek website. This online reference collection remains a valuable resource to anyone studying insects in Minnesota. An untiring naturalist, Haarstad delved deep into the entomological holdings at the University of Minnesota, compiling a "Preliminary Checklist of Minnesota Insects." In preparing the checklist of flower flies included in this book, I've used information from Haarstad and the Cedar Creek collection and checklist.

Certainly, many other people contributed significantly to the knowledge of the flower flies of Minnesota and continue to do so. By collecting flies for the large insect collections at colleges and universities or at the Science Museum of Minnesota, professors, professional biologists, students, and citizen scientists have made important contributions. For instance, both Robert Dana and Ronald Huber have collected many flower flies while pursuing butterflies. And University of Minnesota professor Ralph Holzenthal while documenting the insects visiting his backyard over a number of years by using a malaise trap has added many flower fly records as well.

A malaise trap can be used to collect flower flies.

Mention must be made of a number of important Canadian entomologists who have contributed greatly, generation after generation, to the understanding of North American flower flies. Charles Howard Curran (1894-1972) who described more than two hundred species of *Nearctic flower flies*; Richard Vockeroth (1928-2012) who authored, among other things, *The Flower Flies of the Subfamily Syrphinae of Canada, Alaska and Greenland*; and F. Christian Thompson who currently works at the Smithsonian National Museum of Natural History. The most recent Canadian contribution being the 2019 publication by Jeffrey Skevington et. al. of *Field Guide to the Flower Flies of Northeastern North America*.

The revolution in natural history brought on by the availability of affordable digital cameras can't be underestimated. In conjunction with this change has been the flourishing of citizen science contributions to online databases. Naturalist platforms like eBird, Bugguide.net, AntWiki, OdonataCentral, and iNaturalist (to name but a few) have rapidly expanded and have made it exponentially easier for amateurs and specialists to contribute to our knowledge of the living world. There are also FaceBook groups and other online organizations, such as Diptera.org, where flower fly observations can be discussed. Thus, computer-assisted natural history powers much current work and makes the entrance into specialty areas of the sciences much more attainable to the amateur. It's an exciting time to begin a study of flower flies.

IDENTIFYING FLOWER FLIES

Identifying flower flies can be difficult, especially from photographs. While the ability to identify a fly to species from a photo varies from species to species it depends, also, upon the quality of the photo. It's often helpful to take a number of photos from a variety of angles. At a minimum, try to get images that reveal abdomen patterning and coloration, wing venation, face shape, antennae shape, and legs.

This fly was captured, cooled, then photographed with a macro lens.

Netting flies allows in-hand examination using a loupe. Captured flies can also be cooled and then photographed with a macro lens to reveal difficult to see characteristics. Captured flies can also be killed and pinned and added to a reference collection, allowing for microscopic examination and DNA analysis. A number of flies can only be confidently identified to species after dissection and microscopic examination.

Chapter Two
SYRPHINAE

Syrphinae Genera Summary

Tribe	Genera	Pages
Bacchini	*Baccha, Melanostoma, Platycheirus*	59-68
Paragini	*Paragus*	69-70
Syrphini	*Chrysotoxum, Dasysyrphus, Didea, Doros, Epistrophe, Epistrophella, Eupeodes, Hypocritanus, Lapposyrphus, Leucozona, Megasyrphus, Melangyna, Meligramma, Meliscaeva, Parasyrphus, Pelecinobaccha, Scaeva, Sphaerophoria, Syrphus, Xanthogramma*	71-99
Toxomerini	*Allograpta, Toxomerus*	100-103

Key to the genera of the subfamily Syrphinae

1 Antennae elongated, length equal to or greater than the head width .... ***Chrysotoxum***

1′ Antennae short, the flagellum is ovoid 2

2 Abdomen long and slender, T2 is narrower than scutellum 3

2′ Abdomen tubular or pendant-shaped, equal or wider than scutellum 5

3 Wings hyaline (clear) .... ***Baccha***

3′ Wings with a colored band or colored fore-edge 4

4 Wings with a colored fore-edge .... ***Pelicinobaccha***

4′ Wings with a colored band .... ***Hypocritanus***

5 T1 is well developed (=¾ of T2), extending beyond scutellum .... ***Paragus***

5′ T1 is reduced (<½ of T2), hidden under scutellum 6

6 Scutellum and face are both black 7

6′ Scutellum and face are entirely or partially yellow 8

7 Metasternum normal; Tibia and tarsus of the foreleg of males broadened .... ***Platycheirus***

7′ Metasternum reduced; Tibia and tarsus of the foreleg of males unmodified .... ***Melanostoma***

8 Lower lobe of calypter with long hairs on upper surface (microscope) .... ***Syrphus***
8′ Lower lobe of calypter without long hairs on the upper surface (microscope) 9

9 Rear margin of the eye with a forward-pointing triangular notch ..... ***Toxomerus***
9′ Rear margin of the eye rounded, or smoothly curved 10

10 Anterior anepisternum with long hairs (microscope) 11
10′ Anterior anepisternum without long hairs (microscope) 12

11 Underside of hind coxa with a tuft of two or more hairs (microscope); Abdomen broader than scutellum .... ***Parasyrphus***
11′ Underside of hind coxa without hair (microscope); Abdomen is tubular about equal to scutellum .... ***Meliscaeva***

12 R4+5 vein dips into cell r4+5; Abdomen is broadly oval-shaped; tracheal system visible through abdomen markings 13
12′ R4+5 vein is straight or only slightly sinuous 14

13 Continuous black margin on abdomen .... ***Didea***
13′ Alternating yellow and black margin on abdomen .... ***Megasyrphus***

14 Thorax with lateral stripes extending from humeri to at least the transverse suture 15
14′ Thorax without lateral stripes 19

15 Wide, half-moon spots on segment 4"; Abdomen margin distinctly beaded (microscope) .... ***Epistrophella***
15′ Thin band (complete or broken) or pattern other than spots on segment 4; Abdomen margin not distinctly beaded (microscope) 16

16 Abdomen relatively oval-shaped; Posterior half of Scutellum bright yellow .... ***Xanthogramma***
16′ Abdomen elongate or slightly petiolate 17

17 Relatively large species; Front edge of wings are darkened .... ***Doros***
17′ Small species; Wings clear 18

18 Slender, parallel-sided abdomen; Male with large genital capsule .... ***Sphaerophoria***
18′ Oblique lateral bands on T4; Male genital capsule small, inconspicuous .... ***Allograpta***

19 Metasternum bare (microscope) 20
19′ Metasternum with hair (microscope) 23

20 Eyes with hairs 21
20′ Eyes bare 23

21 Swollen frons; Wing cells extensively bare .... ***Scaeva***
21′ Frons not swollen; Wing cells extensively covered in microtrichia 22

22 Abdomen markings on T3 and T4 do not meet in the middle and are straight .... ***Melangyna*** (in part)
22′ Abdomen markings on T3 and T4 do not meet in the middle and are curved .... ***Meliscaeva***

23 R4+5 loops upward into cell r2+3. Face with black median stripe .... ***Lapposyrphus***
23′ R4+5 straight 24

24 Abdomen with obvious black margin .... ***Eupeodes***
24′ Abdomen without black margin 25

25 Small species with a slender, parallel-shaped abdomen that is rounded without a distinct edge .... ***Melligramma***
25′ Medium-sized species with broader abdomen 26

26 Abdominal markings are straight and do not meet in the middle .... ***Melangyna*** (in part)
26′ Abdominal markings on T3 and T4 meet in the middle forming bands .... ***Epistrophe***

Tribe Bacchini

Tribe Bacchini includes three genera in Minnesota: *Baccha, Melanostoma, Platycheirus*. The latter genus, *Platycheirus*, is one of the most speciose and includes many look-a-like species. Some common characters of the tribe include a black scutellum, a black face (though often with white or silvery pile), an absence of pile on the humeri and a generally concave head.

Genus *Baccha* :: Dainties

The genus *Baccha* includes small flower flies with very slender abdomens, with sections 2 and 3 being narrower than the scutellum. The larvae are predators of aphids. A single species is found in Minnesota.

American Dainty

The American Dainty (***Baccha cognata***) is a small, Nearctic flower fly. It has a long, thin abdomen and clear wings. The abdomen is narrower at the base, about half the width of the scutellum. The larvae feed upon aphids.

Field Notes: A forest dweller, look for it visiting flowers and flying among low vegetation in partially open locations. So far, this fly has been recorded only several times in Minnesota. Interestingly, these observations were made in the far southeast corner of the state and in the far north along the Canadian border. Being such a small fly, perhaps it has simply been overlooked in between.

Similar species: Similar to *Hypocritanus fascipennis* in shape, however that species is larger and has banded wings. Also somewhat similar in size and shape to *Sphegina* and *Neoascia*, but those two species have thickened hind femurs.

Flight Dates: June - August

Flowers: Adults have been observed at *Actaea* (IL).

Range: Eastern

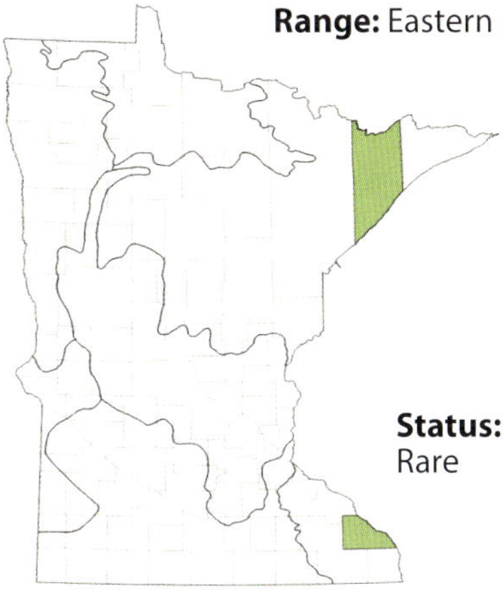

Status: Rare

Genus *Melanostoma* :: Duskyfaces

The genus *Melanostoma* is comprised of small dark flower flies, typically with slender, black abdomens with orange or yellow markings. While numerous species occur in the Palearctic, only a single species is known in the Nearctic. Unlike many other flower flies, they continue to be active in the shade and temperatures as low as 45° F (7° C).

Variable Duskyface

The Variable Duskyface (***Melanostoma mellinum***) is a small, dusky-yellow and black, Holarctic species. Females easily identified by flag-shaped marks on middle segments of the abdomen. Males lack the flattening or broadening of the front legs characteristic of similar Sedgesitter (*Platycheirus*) males. Adults can be found on leaf surfaces where they feed on the pollen of grasses (Poaceae) and other wind-pollinated plants but also visit a variety of flowers. Larvae are entomophagous, mainly aphids.

Female

Field Notes: Flies at temperatures lower than other flower flies, allowing it to get an early start and to fly in the shade.

Similar species: Easily mistaken for flower flies of the closely related genus *Platycheirus*, however, the female abdomen pattern is unique and the males lack the flat, expanded front legs that are characteristic of that genus.

Range: Northern

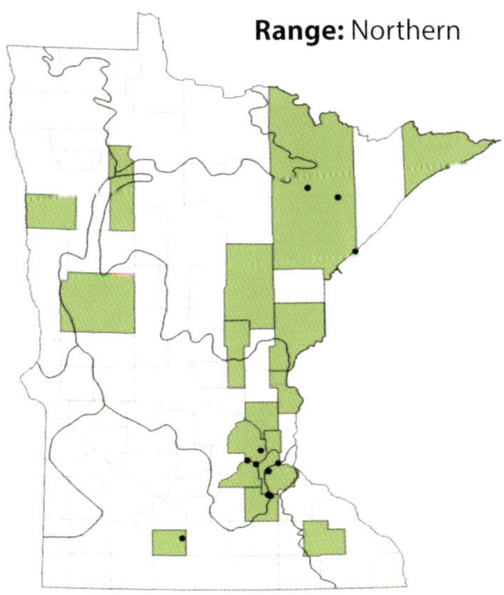

Male

Flight Dates: April – May, September

Flowers: Rue Anemone, Dandelion, Wild Plum, Tansy

Status: Common

Genus *Platycheirus*
Sedgesitters

The genus *Platycheirus* includes small, elongated flies that have dark-colored faces and scutellum. The abdomen is long and either completely black or with orange-yellow, or grey-silver paired markings. Most males have modified front legs. The shape of the tarsi and tibia are useful for identification, as are the location, shape, and color of the hairs on the femur. Females are difficult to identify and some are not separable to species.

The flies usually perch with their wings folded over their abdomens; their abdomen patterns are only visible while in flight.

The numerous species in this genus have been divided into a number of species groups based largely upon the morphology of male legs:

Modified Front Legs of Male Sedgesitters

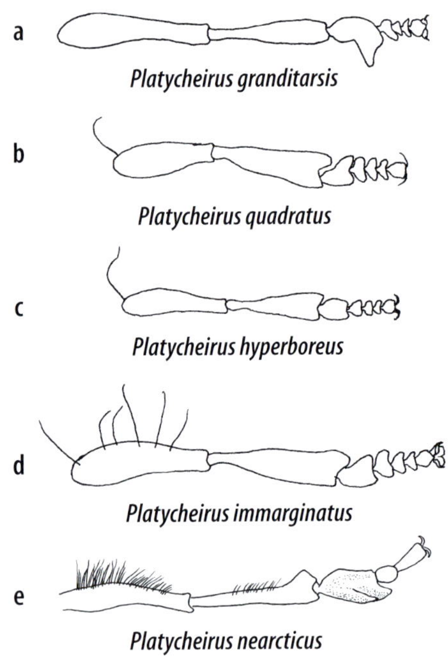

Figures a-d redrawn from Vockeroth, 1992; figure e redrawn from Young et. al., 2016

granditarsis group	large, irregular reddish-orange markings on the abdomen
stegnis group	long black setae on protibia and mesotibia; oblique ripples on facial polinosity
albimanus group	expanded probasitarsus and tibia
pictipes group	unmodified proleg
peltatus group	flattened, posterior setae on profemur

granditarsis group
large, irregular reddish-orange markings on abdomen

Hornhand Sedgesitter

The Hornhand Sedgesitter (***Platycheirus (Pyrophaena) granditarsis***) is a small, reddish-orange and black, Holarctic species. Identifiable by abdomen pattern. The abdomens are slightly pendant-shaped. Males have abdomens that are parallel-sided. Female abdomens are more pendant-shaped with unique yellow markings. Adults can be found on grass and sedge blades at the edge of flowing water, and other wetlands, where they feed on the pollen of grasses (Poaceae) and other wind-pollinated plants. Larvae are entomophagous, mainly aphids.

Male

Range: Northern

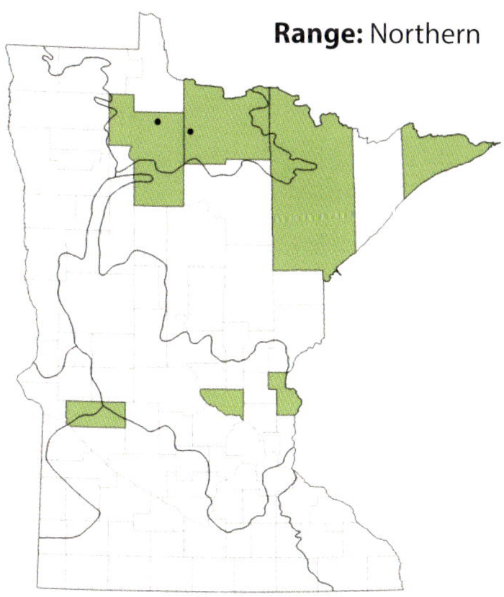

Female

Field Notes: Most common in the northern parts of the state.

Similar species: *Melanostoma mellinum* and *Platycheirus rosarum*.

Flowers: Unknown

Flight Dates: July - August

Status: Uncommon

albimanus group
expanded probasitarsus and tibia

Meadow Sedgesitter

The Meadow Sedgesitter (***Platycheirus quadratus***) is a small, dusky-yellow and black, Nearctic species. Males can be identified by abdomen pattern and widened front feet, tibia and tarsus (*see drawing, p. 62*). Females are not easily separable from other Sedgesitter (*Platycheirus*) species. Adults are commonly found on grass and sedge blades at the edge of ponds and wetlands where they feed on pollen grains.

Male

Field Notes: This is the most common *Platycheirus* species in Minnesota, abundant statewide.

Similar species: *Melanostoma mellinum* and other *Platycheirus* species.

Flowers: Carex

Flight Dates: May - October

Status: Common

Range: Southern

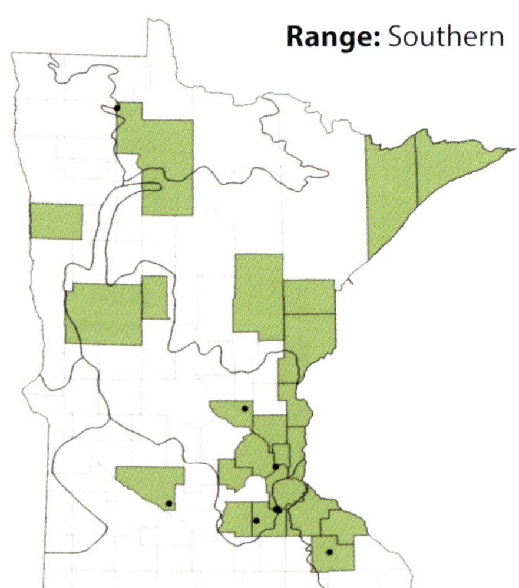

Pearly Sedgesitter

The Pearly Sedgesitter (***Platycheirus hyperboreus***) is a small, pearly-yellow and black, Nearctic species. The yellow abdominal marks of both males and females are overlaid with polinosity which lends a pearly sheen to the color. Males can be identified by abdomen pattern, yellow markings with wide-margins of black and by the shape of their widened front feet, tibia and tarsus (*see drawing, p. 62*); the probasitarsus is much more regular in shape than that of *P. quadratus*) Females look similar to males but are not easily separable from other Sedgesitter (*Platycheirus*) species. Adults are most likely to be found on grass and sedge blades at the edge of ponds and wetlands where they feed on pollen grains.

Female

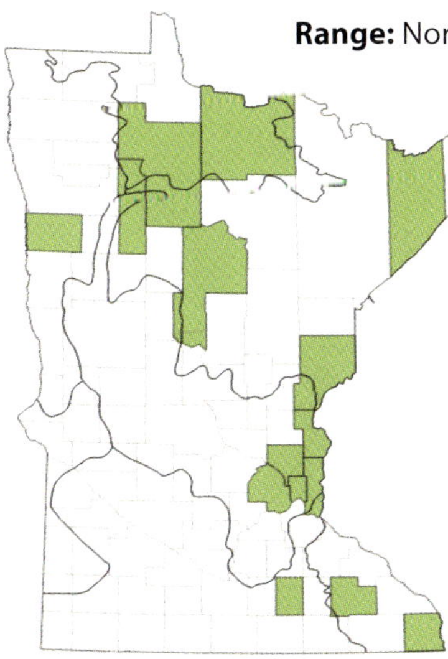

Range: Northern

Status: Uncommon

Field Notes: Expected in a wide-variety of wetlands, however, we have no modern photos of this species.

Similar species: *Melanostoma mellinum* and other *Platycheirus* species.

Flowers: Several dozen of these flies are pinned with associated grass seed heads (reed canary?) on which they were presumably found. All these specimens date from 1969 in St. Cloud in August.

Flight Dates: June - September

Blackspine Sedgesitter

The Blackspine Sedgesitter (***Platycheirus scambus***) is a small, dusky-yellow and black, Nearctic species. Males can be identified by abdomen pattern and widened front feet, tibia and tarsus (*see drawing, p. 62*). Females are not easily separable from other Sedgesitter (*Platycheirus*) species. Adults are commonly found on grass and sedge blades at the edge of ponds and wetlands where they feed on pollen grains.

Male

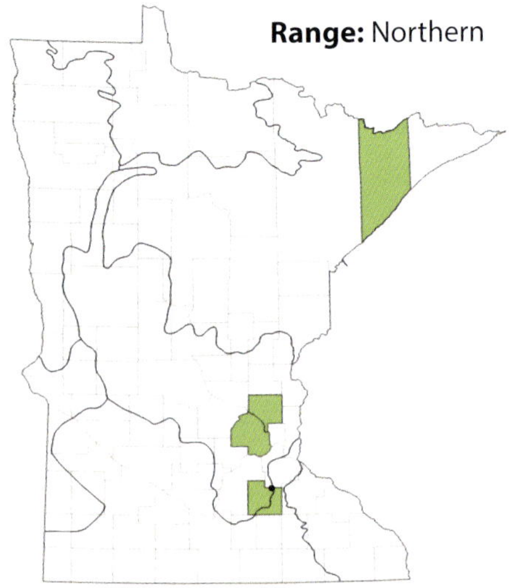

Range: Northern

Flowers: n/a

Flight Dates: June

Status: Uncommon

Field Notes: Most recent observation found in Reed Canary choked drainage ditch.

Similar species: Most similar to the much more common *P. quadratus*. Microscope examination of prolegs needed for positive identification.

peltatus group: expanded probasitarsus and tibia expanded at apex only

Nearctic Broadhand Sedgesitter

The Nearctic Broadhand Sedgesitter (***Platycheirus nearcticus***) is a small, dusky-yellow and black, Nearctic species. Males can be identified by abdomen pattern and widened front feet, tibia and tarsus (*see drawing, p. 62*). Females are not easily separable from other Sedgesitter (*Platycheirus*) species. Adults are commonly found on grass and sedge blades at the edge of ponds and wetlands where they feed on pollen grains.

Male

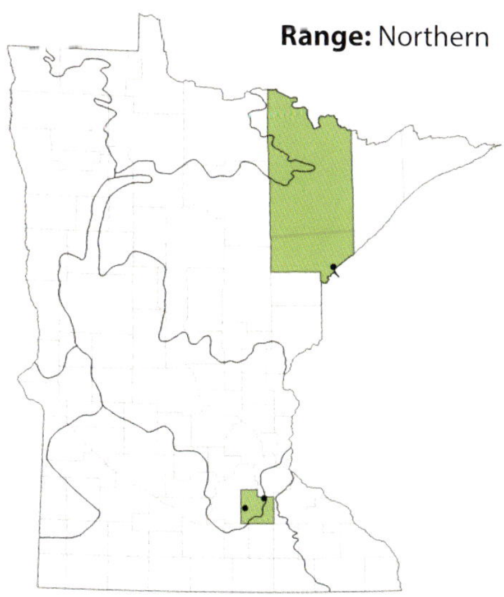

Range: Northern

Field Notes: Has been found visiting hilltop flowers.

Similar species: Most similar to the much more common *P. quadratus. Microscope examination* of prolegs needed for positive identification.

Flowers: Adults have been found to visit cultivated *Phlox*, and have been reported to visit *Ranunculus*.

Flight Dates: June

Status: Rare

Additional Minnesota Bacchini Tribe Species Not Included in the Guide

stegnis group: long black setae on protibia and mesotibia; oblique ripples on facial polinosity

Platycheirus confusus, Confusing Sedgesitter

Platycheirus obscurus, Eastern Forest Sedgesitter

granditarsis group

Platycheirus (Pyrophaena) rosarum, Fourspot Sedgesitter

albimanus group

Platycheirus angustatus, Delicate Sedgesitter

Platycheirus clypeatus, * Smoky-winged Sedgesitter

Platycheirus immarginatus, Comb-legged Sedgesitter

Platycheirus modestus, Yellow Sedgesitter

Platycheirus nodosus, * Twospear Sedgesitter

Platycheirus scutatus, * Many-tufted Sedgesitter

Platycheirus thompsoni, Thomspson's Sedgesitter

Platycheirus varipes, Silver Sedgesitter

pictipes group: unmodified proleg

Platycheirus luteipennis, Coppery Sedgesitter

Platycheirus pictipes, Cobalt Sedgesitter

Tribe Paragini

Tribe Paragini includes a single genus in Minnesota: *Paragus*. The flies in this genus are very small, usually less than 5 mm (0.2 in) in length. They are generally black-bodied and have yellow faces.

Genus *Paragus* :: Grass Skimmers

Tiny, black flower flies that keep close to the ground. The first segment of the abdomen is wider than in most other Syrphidae and is visible. The genus is divided into two subgenera: *Paragus* and *Pandasyophthalmus*.

Black-backed Grass Skimmer

The Black-backed Grass Skimmer (***Paragus (Pandasyophthalmus) haemorrhous***) is a tiny, Nearctic flower fly. The scutellum is entirely black. The face is yellow with a black, medial stripe. The eyes are uniformly pilose, without visible stripes. The end of the abdomen, from segment 3, is a deep reddish-orange color.

Range: Southern

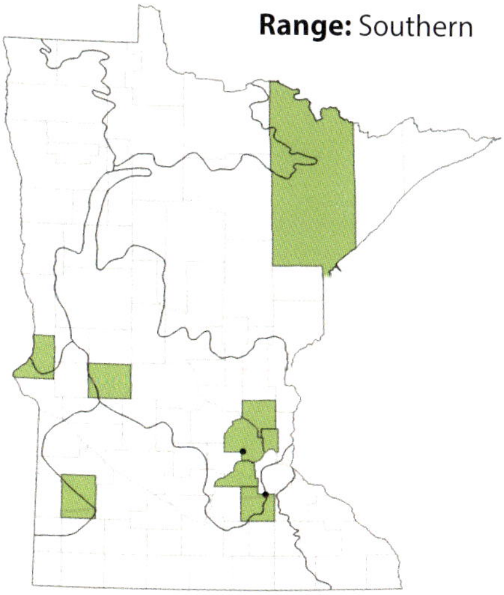

Field Notes: This is the only Minnesota species identifiable to species; all the others require dissection.

Similar species: Most similar to other *Paragus* species. Could be mistaken for *Eumerus*.

Status: Uncommon

Flowers: Photographed on leafy spurge and goatsbeard in Minnesota. Known to visit numerous flowering plants.

Flight Dates: June - July

Grass Skimmers

The Grass Skimmers (***Paragus (Paragus)*** spp.) are tiny, Nearctic flower flies. The scutellum is edged with yellow. The face is yellow with or without a black, medial stripe. Eyes have several stripes of pile. Dissection is required to separate species.

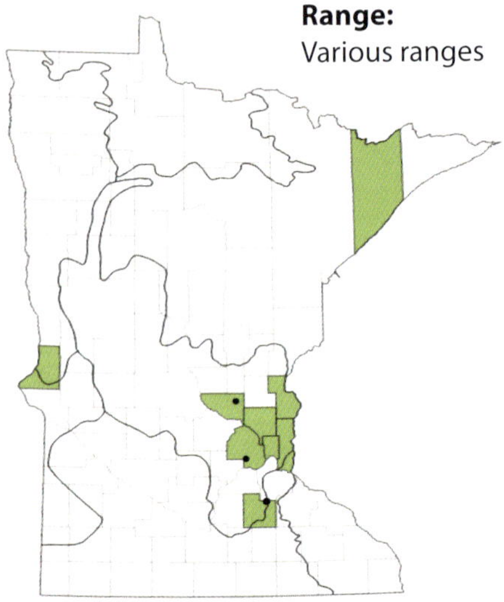

Range:
Various ranges

Field Notes: Common, but easily missed because of size. Look for these flower flies perched on low vegetation at the edge of open areas and trails.

Similar species: Most similar to other *Paragus* species. Could be mistaken for *Eumerus*.

Status: Common

Flowers: While often found perched at the tip of grass blades and other low vegetation, they have been observed visiting white snakeroot in Minnesota.

Flight Dates: May - September

Tribe Syrphini

Tribe Syrphini includes numerous genera in Minnesota: *Chrysotoxum, Dasysyrphus, Didea, Epistrophe, Epistrophella, Eupeodes, Hypocritanus, Lapposyrphus, Leucozona, Megasyrphus, Melangyna, Meligramma, Meliscaeva, Parasyrphus, Pelecinobaccha, Scaeva, Sphaerophoria, Syrphus, Xanthogramma.*

Genus *Chrysotoxum* :: Meadow Flies

The flies in this genus are large, convincing wasp mimics. The wide abdomens have distinctive, arched yellow markings. The antennae are very long. The wings are colored along the front margin to mimic the folded wings of vespid wasps. Larvae are believed to be associated with root aphids tended by ants.

The genus is easily recognized but the species are very difficult to separate, especially in photographs.

Narrow-banded Meadow Fly

The Narrow-banded Meadow Fly (***Chrysotoxum derivatum***) is a large, Nearctic flower fly. Segment 1 of the antenna is longer than segment 2. Abdominal bands slightly narrower than those of *C. plumeum*. The basal half of the metafemur is black.

Field Notes: Associated with the boreal forests of northern Minnesota.

Similar species: Difficult to separate from other species in the genus without microscope examination.

Flowers: Only a single floral record is reported elsewhere and that is goldenrod.

Flight Dates: August

Status: Rare

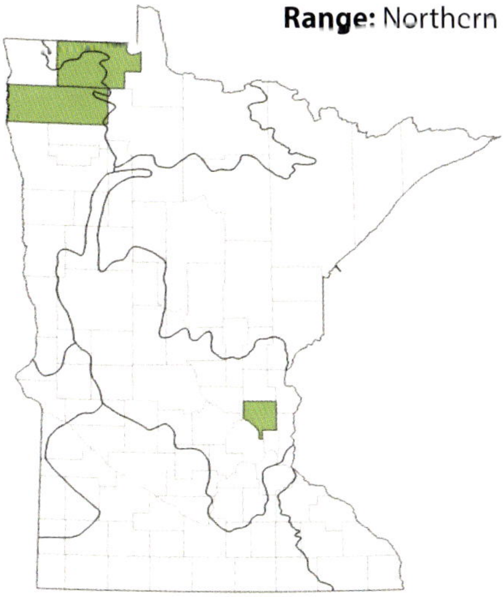

Broad-banded Meadow Fly

The Broad-banded Meadow Fly (***Chrysotoxum plumeum***) is a large, Nearctic flower fly. Segment 1 of the antenna is longer than segment 2. Abdominal bands slightly wider than those of *C. derivatum*. Metafemur mostly yellow with small band of black near the base. Pile on abdominal segments 2 and 3 is mostly yellow.

Female

Range: Northern

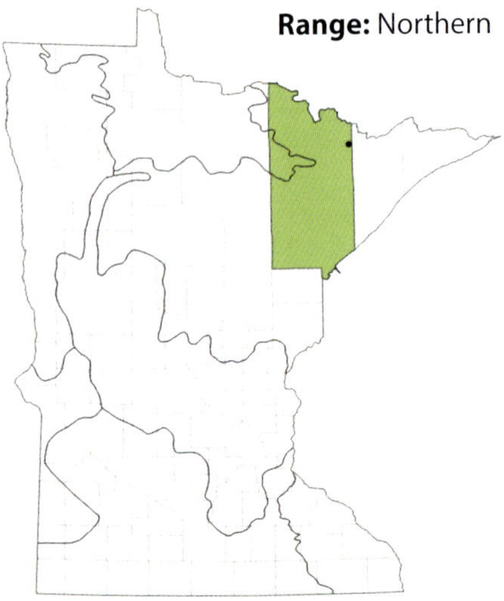

Field Notes: Associated with the boreal forests of northern Minnesota.

Similar species: Difficult to separate from other species in the genus without microscope examination.

Status: Uncommon

Flowers: Very few flower records exist other than the reported goldenrod (*Solidago*).

Flight Dates: June

Yellow-throated Meadow Fly

The Yellow-throated Meadow Fly (***Chrysotoxum pubescens***) is a large, Nearctic species. It is the only species of this genus in our area to have a yellow proepimeron. The antennae are long, with segment 3 about equal in length to the combined length of segments 1 and 2 (note also that segment 2 is nearly twice the length of segment 1). The wings are lightly colored along the front margin. The scutellum is yellow around the edges and black in the center. Abdomen pattern similar to other species, however, the Yellow-throated Meadow Fly has orange coloration at the posterior edges and corners of the abdominal segments. The legs are mostly yellow in color. The femur of the hind leg has an orange band. The tarsi are also orange in color.

Field Notes: Larger than other species, but this is difficult to judge in the field or in photographs.

Similar species: Similar in appearance to other meadow flies, but the only species with yellow proepimeron, making it the only easily identified species in this difficult group of look-a-likes.

Flowers: Gray Dogwood in Minnesota

Flight Dates: June

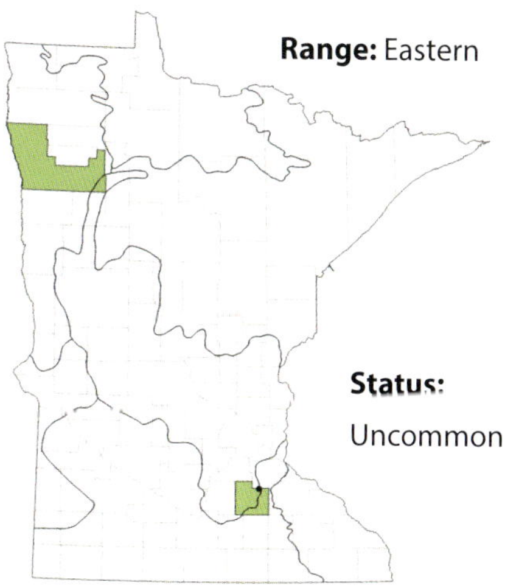

Range: Eastern

Status: Uncommon

Additional Minnesota Syrphini Tribe Species Not Included in the Guide

Chrysotoxum flavifrons, Blackshield Meadow Fly: Northern species with long flagellum and long pile on anterior anepisterum.

Genus *Dasysyrphus* :: Conifer Flies

The flies in this genus have straight or lunate markings on the abdomen that usually do not meet in the middle. Their wings have a distinct, elongate black stigma. The larvae feed on aphids that live in treetops, in many cases on conifers. Three species are possible in Minnesota.

Confusing Conifer Fly

The Confusing Conifer Fly (***Dasysyrphus intrudens***) is a medium-sized, Nearctic species. It has a wide abdomen with paired markings on abdomen segments 2 - 4, usually white or pale yellow in color. The markings on segments 3 and 4 are constricted at the center, with those on segment 4 even divided in half on occasion. Larvae feed on aphids.

Field Notes: The species concept includes several cryptic species, distinct genetically but nearly identical morphologically.

Similar species: Most similar to *Dasysyrphus venustus*, differing only by the degree to which the abdominal markings on segments 3 and 4 are constricted in the middle.

Flowers: No flower associations are known from Minnesota, but the flies have been recorded visiting a wide range of flowers elsewhere in their range.

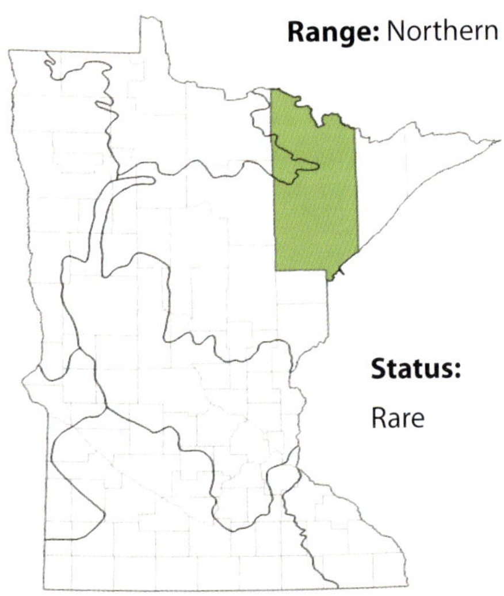

Range: Northern

Status: Rare

Flight Dates: June - August

Genus *Didea* :: Lucents

The flies in this genus are quite distinct and easily recognized. They have bold abdominal markings, visible tracheal system, and the R4+5 vein loops downward, which is an unusual trait among the syrphines. There are only two Nearctic species.

Undivided Lucent

The Undivided Lucent (***Didea fuscipes***) is a medium-sized, Nearctic species. The pattern on the rather wide and oval-shaped abdomen is bold and distinct and can vary in color from yellow to lime-green to a lucent blue. Vein R4+5 loops downward into cell r4+5, reminiscent of eristaline wing venation.

Male

Field Notes: The tracheal system is visible through the translucent markings on the abdomen, especially noticeable in live specimens.

Similar species: Most similar to *Megasyrphus.*

Flowers: n/a

Status: Rare. Only two specimens known for Minnesota.

Flight Dates: June

Range: Southern

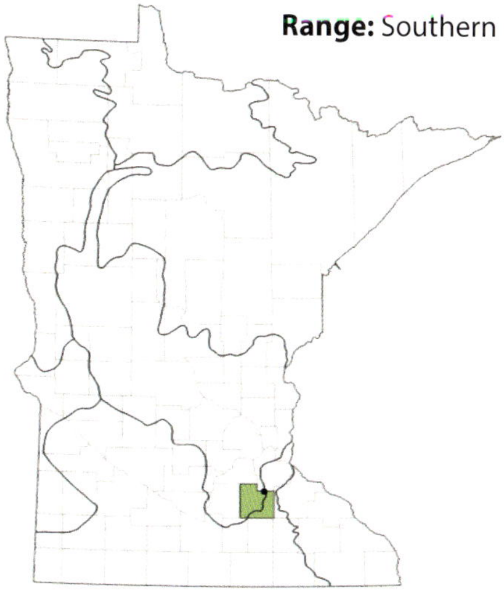

Genus *Doros* :: Potterflies

Only one species of this genus is found in North America. They have a tubular-shaped abdomen with thin yellow bands and are convincing mimics of vespid wasps, especially mason wasps such as *Ancistrocerus*. The larvae are thought to be predatory of root aphids associated with ant nests.

Canadian Potterfly

The Canadian Potterfly (***Doros aequalis***) is a medium-sized, Nearctic species. It has a tubular-shaped body with thin yellow bands. It also has thin yellow markings along the top edges of the thorax. The leading edge of the wing is darkened to mimic the thin, dark wings of the vespid wasps. The larvae, like those of the closely related *Xanthogramma*, are thought to prey upon root aphids associated with ant nests.

Field Notes: A known hilltopper. Open areas on hilltops in the northern part of the state would be good places to look for this spectacular, and rare, fly. In the U.K., the sister species, *Doras profiles*, is so infrequently encountered that it's been give the name Phantom Hoverfly.

Range: Northern, with several disjunct southern populations

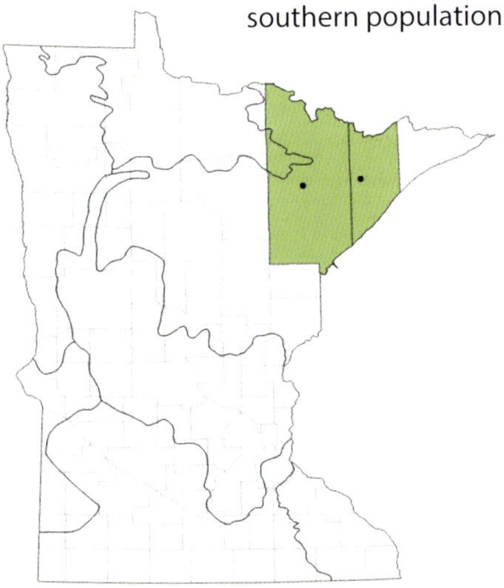

Similar species: Most similar to the wasps it mimics. Not likely to be confused for any other flower fly, with the possible exception of *Spilomyia sayi* which also mimics potter wasps.

Flowers: No floral records are available for Minnesota. Elsewhere in its range, these flies have been observed on false Solomon's seal, wild rose, anemone, dogwood, and meadow rue.

Flight Dates: June

Status: Rare

Genus *Epistrophe* :: Smoothtails

The flies in this genus are typical syrphines: yellow-and-black banded abdomens, yellow faces, and yellow scutellum. They differ from the similar-looking genus *Syrphus* by not having long setae on the upper surface of the lower calypter.

Black-horned Smoothtail

The Black-horned Smoothtail (***Epistrophe grossulariae***) is a medium-sized, Holarctic flower fly. Yellow scutellum. Complete yellow bands on abdomen segments 3 and 4. Separated from other *Epistrophe* species by all-black antennae and the black facial triangle above the antennae. The larvae feed upon aphids on trees but are also known to feed on aphids on crops of *Brassica* and Wheat.

Range: Northern

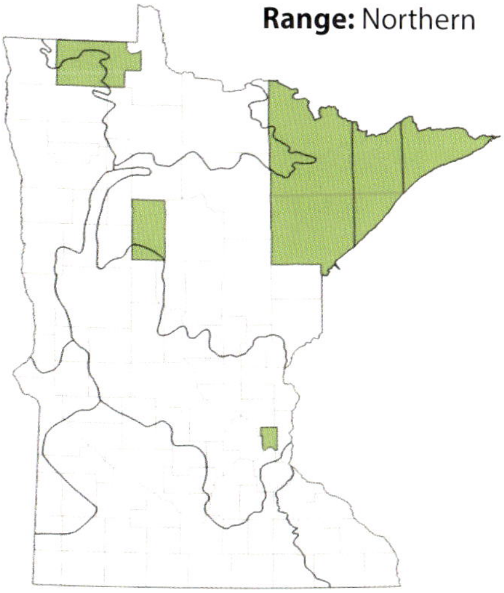

Field Notes: The larvae of this scarce flower fly are known for their prolonged diapause which can last three or more years.

Similar species: Can be confused with other yellow-and-black syrphines, especially *Syrphus* and *Parasyrphus*.

Flowers: No floral records for Minnesota. Adults, elsewhere, visit a wide variety of flowers.

Flight Dates: July - September

Status: Common

Genus *Epistrophella* :: Smoothtails

The two species of *Epistrophella*, one Palearctic and one Nearctic, were previously placed as a subgenus of *Epistrophe*.

Slender Smoothtail

The Slender Smoothtail (***Epistrophella emarginata***) is a small, Nearctic species. The thorax has distinct lateral stripes that are sometimes rather hard to see. This species has a somewhat slender abdomen with yellow markings. The typical pattern is a complete band on segment 3 and a divided band on segments 2 and 4, but this pattern varies greatly. The larvae feed on aphids and can be identified by the triple points at the edge of each segment.

Field Notes: The abdomen pattern is highly variable, resulting in some taxonomic confusion that remains to be sorted out. Many synonyms exist.

Similar species: Can be confused with other yellow-and-black syrphines.

Flowers: No floral records for Minnesota. Adults, elsewhere, visit a wide range of flowers, including milkweeds, willows, and asters.

Flight Dates: September

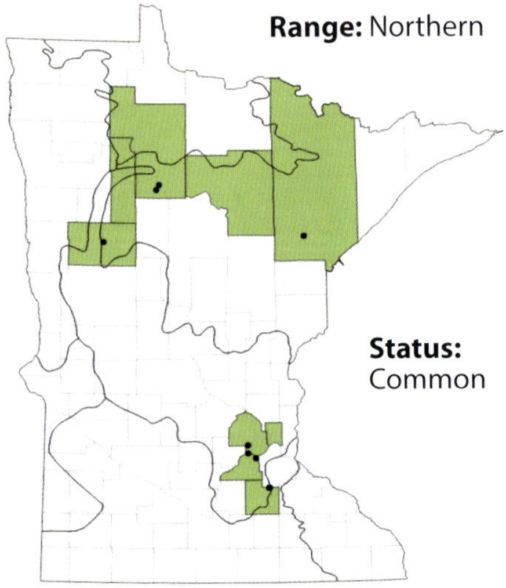

Range: Northern

Status: Common

Genus *Eupeodes* :: Aphideaters

The flies of the genus *Eupeodes* are typical yellow-and-black banded syrphines. They are recognized by their black-margined abdomen that has a crimped look when examined closely. They also have a pilose metasternum. Twenty-four North American species are described.

Large-tailed Aphideater

The Large-tailed Aphideater (***Eupeodes volucris***) is a small, Nearctic species. Males are easily recognized by the cylindrical protrusion of the genitalia. The scutellum usually has a white border along the posterior edge. The cells of the wings are extensively bare at the base. The paired markings on abdomen sections 2, 3, and 4 are curved on the latter two sections. Abdomen markings are often white but occasionally yellow. Female sternite sections 4 and 5 are orange-colored, without black marks.

Range: Western

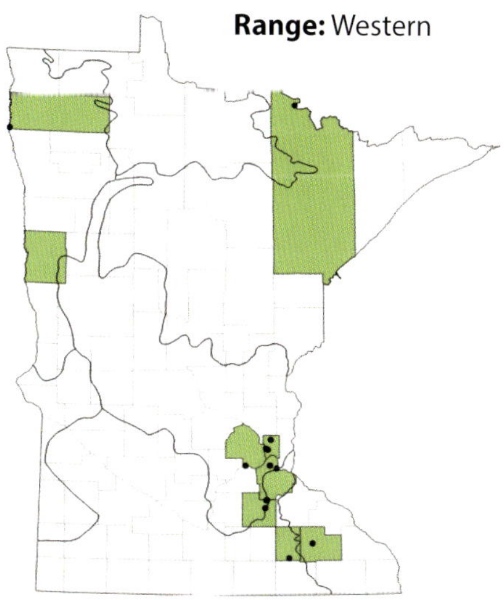

Field Notes: Also known by the common name Bird Hoverfly. A known migrant. Most likely to be observed in the western parts of the state. Often observed resting on vegetation adjacent mud puddles.

Similar species: Most similar to other *Eupeodes* species, but also could be confused with *Scaeva* and *Lapposyrphus*.

Flowers: A generalist that visits a vast number of flowering plants, fond of invasive umbellifers such as Queen Anne's lace and wild parsnip.

Status: Abundant

Flight Dates: May - September

Long-tailed Aphideater

The Long-tailed Aphideater (***Eupeodes americanus***) is a small, Nearctic species. Males are only separable from the identically colored but very rare species (*Eupeodes pomus*) by the shape and length of the genitalia—*E. americanus surstylus* is long and somewhat lyre-shaped while those of *E. pomus* are short and stout. Females cannot be separated morphologically, only by DNA analysis. The scutellum is yellow. The yellow bands on abdominal sections 3 and 4 are complete and are not notched in the center. The bands do not cross the abdomen margin.

Range: Northern

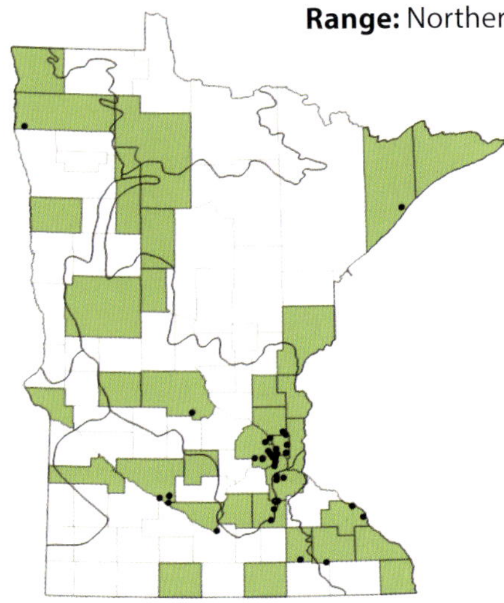

Field Notes: Horace Telford reports this flower fly as the most abundant species in Minnesota in 1932; it still may be. Late in the year, large numbers may be observed along damp gravel trails. These flies frequently land on the bare skin on the hands and arms of observers, which may be the cause of the "mistaken" vernacular name "sweat bee."

Similar species: Very frequently confused with similarly-patterned *Syrphus* species.

Flowers: Adults visit a wide range of flowering plants.

Status: Abundant

Flight Dates: May - October

Bare-winged Aphideater

The Bare-winged Aphideater (***Eupeodes perplexus***) is a medium-sized, Nearctic species. Basal wing cells are extensively bare. Male abdomens are wide and pendant-shaped and their genitalia is not visible from above. The underside of the female abdomen has wide, black bands on the sternites.

Field Notes: Noticeably larger and more black than the much more common species of aphideater.

Similar species: Most similar to *Eupeodes volucris* but larger and more dark.

Flowers: Aster in Minnesota. Elsewhere, they are reported to visit other composites.

Flight Dates: September

Status: Uncommon

Range: Northern

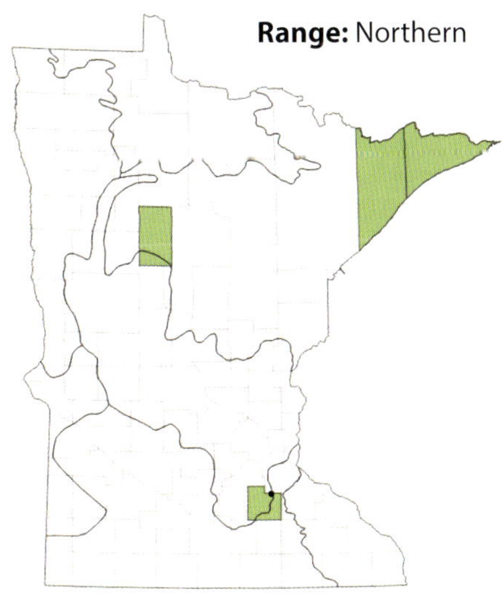

Additional Minnesota Species Not Included in the Guide

Eupeodes pomus; Eupeodes latifasciatus: alula completely trichose, other Minnesota species have bare area.

Genus *Lapposyrphus* :: Aphideaters

The single Nearctic species in this genus was previously classified as a member of *Eupeodes*. The R4+5 vein curves upward, much more markedly than in *Eupeodes*.

Common Loopwing Aphideater

The Common Loopwing Aphideater (***Lapposyrphus lapponicus***) is a medium-sized, Holarctic flower fly. Abdomen sections 3 and 4 have a pair of curved, light-yellow or white markings. The R4+5 vein is curved upwards into r2+3 cell (*Eupeodes* curves upwards as well, but later and less and is matched by a slight downward curve in R2+3 vein). Scutellum solidly colored, without white color to the hind edge as in *Eupeodes volucris*.

Field Notes: This species is a known migrant. As with most Holarctic species, the main distribution of the Common Loopwing Aphideater is the north of Minnesota.

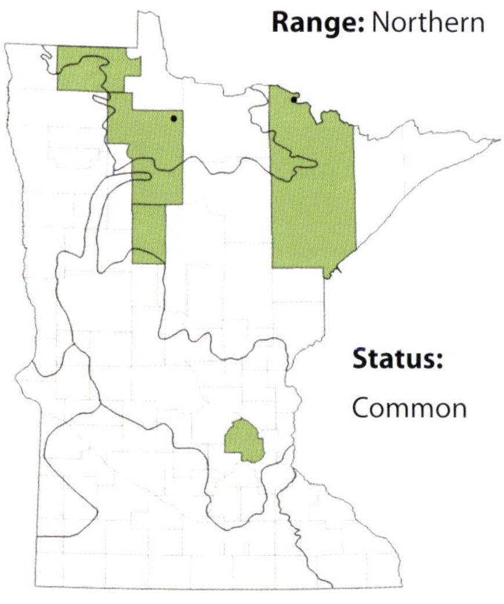

Range: Northern

Status: Common

Similar species: The abdomen pattern is most similar to *Eupeodes volucris* and *Scaeva affinis*. Subtle differences in wing venation and trichosity of wing cells separate these species.

Flowers: Adults have been recorded at a large variety of flowers, especially boreal and bog species of flowers.

Flight Dates: June - September

Genus *Leucozona* :: Whitebelts

The flies of the genus *Leucozona* have extensive white markings on the second abdomen segment which is continuous with thin white markings on the edge of the third segment. The remainder of the abdomen is black. Larvae prey upon aphids. Currently, there are no known records for Minnesota.

American Whitebelt

The American Whitebelt (***Leucozona americana***) is a large, Nearctic flower fly. It has extensive white markings on the second abdomen segment which is continuous with thin white markings on the edge of the third segment. The remainder of the abdomen is black. Larvae prey upon aphids.

Field Notes: n/a

Range: Northern

Similar species: This species should not be confused with other flower flies. That being said it could possibly be mistaken for a drone fly (genus *Eristalis*) though it differs in coloration and wing venation.

Flowers: Adults are reported to visit dogwood and sweet clover.

Flight Dates: n/a

Status: Hypothetical

Genus *Megasyrphus* :: Gossamers

There are two Nearctic species in this genus. The R4+5 vein curves downward, though not as much as is typical of many eristalines.

Black-legged Gossamer

The Black-legged Gossamer (***Megasyrphus laxus***) is a medium-large, Nearctic flower fly. The R4+5 vein curves downward, though not as much as is typical of many eristalines. Like *Didea*, the color of the abdominal marks range from yellow to translucent green, with visible trachea. The bands on abdomen sections 3 and 4 are roundly notched in the center. Larvae are thought to prey on aphids in the forest canopy.

Field Notes: n/a

Similar species: Can easily be mistaken for other yellow-and-black syrphines.

Flowers: The single Minnesota photo record has this fly visiting wild rose. Elsewhere it visits a wide variety of flowers.

Flight Dates: June - October

Status: Uncommon

Range: Northern, with few mid-continent records.

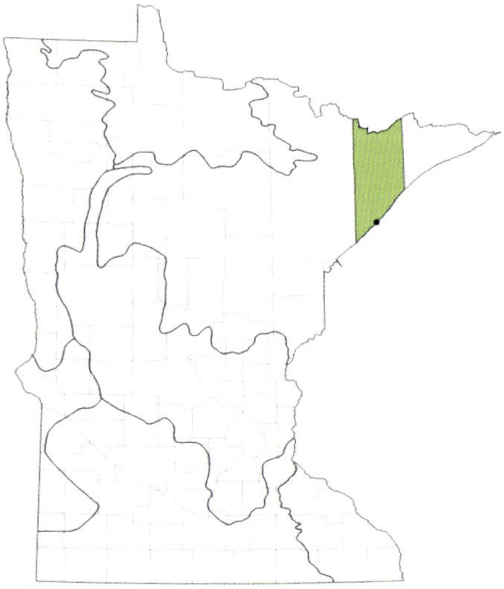

Genus *Melangyna* :: Halfbands

The flies of the genus *Melangyna* have unmargined abdomens. The marks on the abdomen tend to be uncurved and somewhat rectangular in shape. So far, two species are known from Minnesota but at least one more species is expected.

Large-spotted Halfband

The Large-spotted Halfband (***Melangyna fisherii***) is a medium-sized, Nearctic species. The abdominal markings of the female are white or pale-yellow, with those on segment 2 being the largest and almost square in shape. The male pattern is less distinct.

Field Notes: Look for this rare species in mixed-boreal forest and bog habitats.

Range: Northern

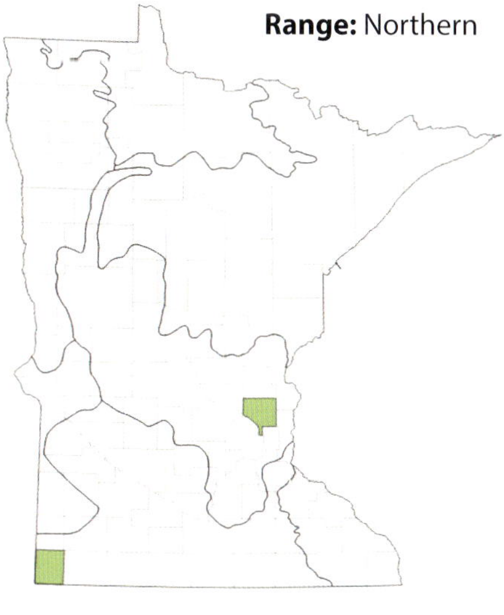

Similar species: Males are difficult to separate from others of the same genus and other similar-colored syrphines.

Flowers: No floral records for Minnesota. Adults, elsewhere, have been recorded at *Tragopogon* and numerous flowers of the family Apiaceae.

Flight Dates: June - October

Status: Rare

Hair-eyed Halfband

The Hair-eyed Halfband (***Melangyna lasiophthalma***) is a medium-sized, Holarctic species. The abdominal markings of the female are white or pale-yellow, with those on segment 2 being the largest and almost square in shape. The abdomen is unmargined. The male pattern is similar but less distinct. Females have a pollinose band on the frons above the antennae. Wing cells c and bm are bare basally.

Field Notes: Flies very early in the season. Look for this rare species in mixed-boreal forest and bog habitats.

Similar species: Males are difficult to separate from others of the same genus and other similar-colored syrphines.

Flowers: No floral records for Minnesota. Adults, elsewhere, have been recorded at *Tragopogon* and numerous flowers of the family Apiaceae.

Flight Dates: March - May

Status: Rare

Range: Northern

Genus *Meligramma* :: Roundtails

The small flies of the genus *Meligramma* have slender, tubular abdomens with rather circumspect markings. Only three species are found in North America and only one of those is known from Minnesota.

Variable Roundtail

The Variable Roundtail (***Meligramma triangulifera***) is a small, Nearctic species. Abdominal sections 2-4 have paired yellow spots, these are thin, taking up about a 1/3 of the segment, and, sometimes, the spots connect on segments 3 and 4 to form thin bands. The frons of the females above the antennae is shiny black.

Field Notes: Look for this rare species in mixed-boreal forest and bog habitats.

Similar species: Because of their abdomen shape, these flies could be mistaken for those of genus *Platycheirus* but the yellow scutellum and yellow face differentiates them.

Flowers: No Minnesota floral records. Adults, elsewhere, report to visit a variety of flowering plants.

Flight Dates: June - October

Range: Northern

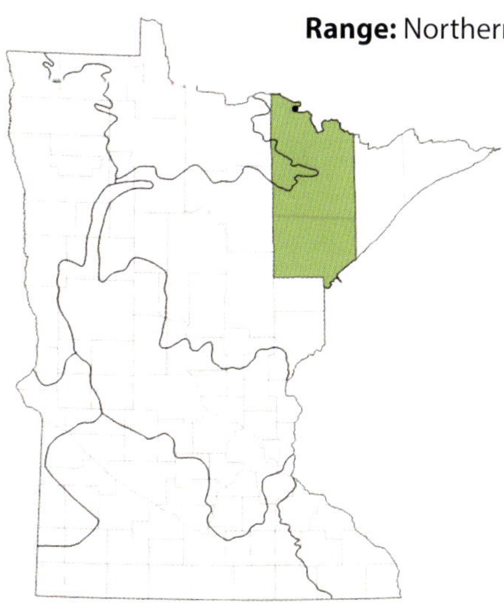

Status: Rare

Genus *Meliscaeva* :: Thintails

The small flies of the genus *Meliscaeva* have somewhat slender abdomens with yellow-and-black markings and wings that are slightly longer than the abdomen. While twenty-five species occur in North America only a single species is known from Minnesota.

Common Thintail

The Common Thintail (***Meliscaeva cinctella***) is a small, Holarctic flower fly. The abdomen is parallel-sided. Segment 2 of the abdomen has wide yellow marks that don't meet in the middle. Segments 3 and 4 with wide, complete bands. Faint lateral stripes on thorax. Larvae fead on aphids.

Field Notes: Common Thintails are known to visit a wide variety of habitats.

Range: Northern

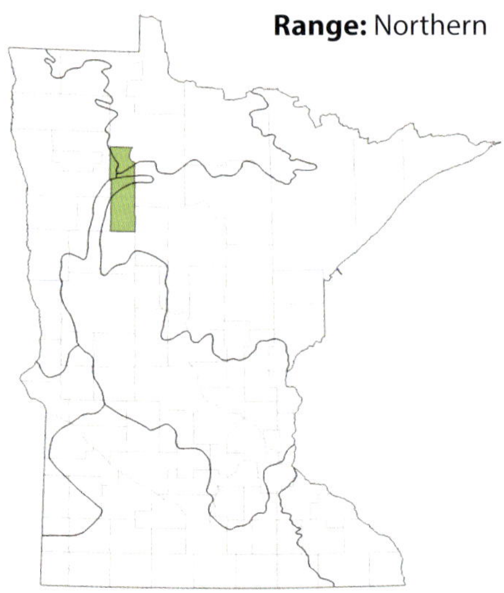

Similar species: Most similar to other black-and-yellow syrphinines, such as *Parasyrphus*.

Flowers: No foral records for Minnesota. Adults, elsewhere, are known to visit forget-me-not and goldenrod.

Flight Dates: June - September

Status: Rare

Genus *Hypocritanus* :: Hover Flies

The flies of the genus *Hypocritanus* are small to medium-sized, with long, slender abdomens, though the forms and colors vary greatly in the tropics of Central and South America. Several species in this genus have adapted to live in the aquatic, micro-habitats found in bromeliad tanks. These same larvae use venom to subdue their prey. A single species is known from Minnesota. This genus was recently split from *Ocyptamus*.

Eastern Band-winged Hover Fly

The Eastern Band-winged Hover Fly (***Hypocritanus fascipennis***) is a small, Nearctic flower fly with a long, slender abdomen. Wings have a dark, triangular-shaped band near the stigma. Larvae feed on aphids.

Field Notes: At certain times of the year, males can be found hovering for extended periods of time in sunlit clearings at heights of ten feet and higher. When perched with wings spread, these flies look like tiny, two-winged dragonflies.

Similar species: Most similar to *Baccha cognata* (p. 59) but that species is even smaller, and has more of a wasp-waist and doesn't have the dark banding of the wings.

Flowers: Minnesota floral records include wood sorel, goldenrod, amaranth, mustard, ragwort, and touch-me-not.

Flight Dates: May - September, with most observations being made in August.

Status: Common

Range: Southern

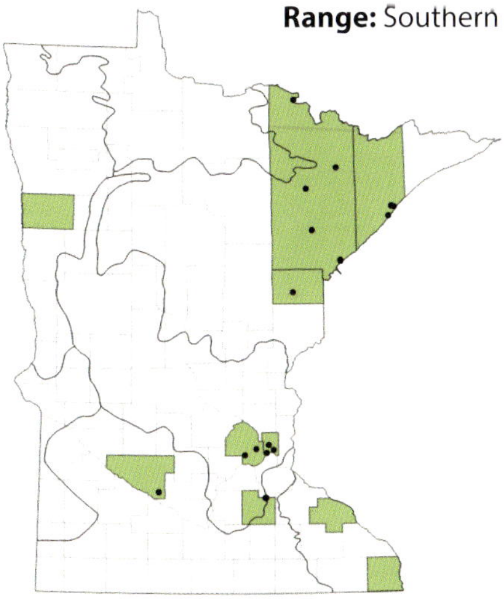

Genus *Parasyrphus* :: Bristlesides

The flies of the genus *Parasyrphus* differ from the very similar genus *Syrphus* by having pile on the anterior anepisternum and a tuft of pile on the posteromedial apical angle of the metacoxa. Most have dark hind legs.

Vockeroth's Bristleside

The Vockeroth's Bristleside (***Parasyrphus vockerothi***) is a medium-sized, Nearctic flower fly. Sternites with bold, black bands. Male frons extensively pollinose, with shiny areas just above the antennae. Wings densely microtrichose.

Field Notes: The one recorded observation for Minnesota was found basking on sunlit Arborvitae (*Thuja*).

Range: Northern

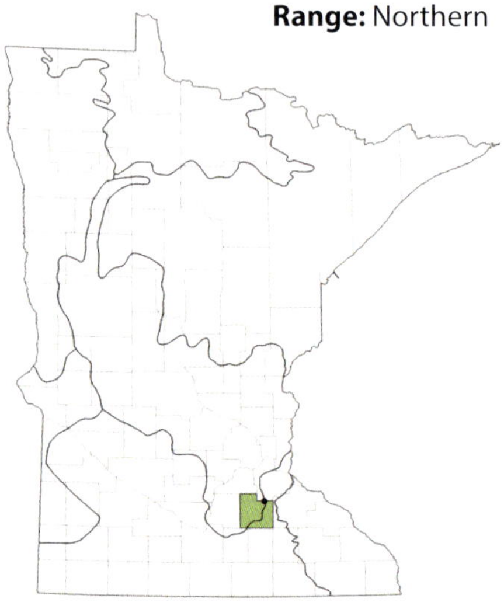

Similar species: Most similar to other yellow-and-black syrphines with complete yellow bands, such as *Eupeodes* and *Syrphus*.

Flowers: Adults have been reported to visit *Caltha, Chamaedaphne*, and *Malus*.

Flight Dates: May

Status: Rare

Additional Parasyrphus Minnesota Species Not Included in the Guide

Parasyrphus genualis (Northern); ***Parasyrphus nigritarsis*** * (Northern)

Genus *Pelecinobaccha* :: Cobalt Hover Flies

The genus *Pelecinobaccha* was only recently split from the genus *Ocyptamus*. Of the more than fifty species in this Neotropical genus, only one species is found north of Mexico. The flies have narrow abdomens and the larvae are known to feed on scale insects (Coccoidea).

Cobalt Hover Fly

The Cobalt Hover Fly (***Pelecinobaccha costata***) is a medium-sized, Nearctic flower fly. This fly has a slender, wasp-like abdomen with pale triangular spots on the sides. The wings are darkened on the leading edges. Larvae feed on scale insects on trees.

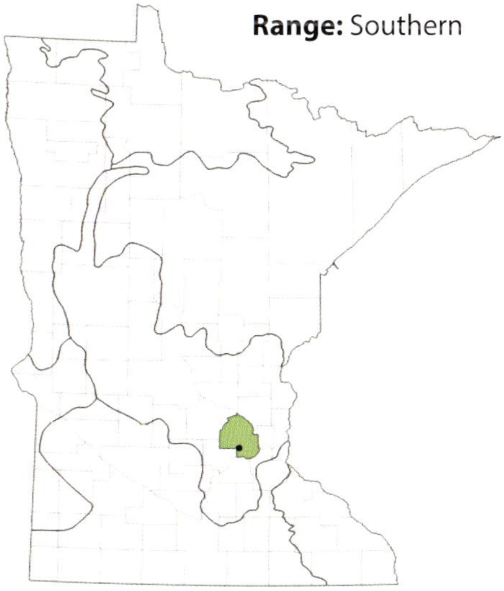

Range: Southern

Field Notes: Observed for the first time in Minnesota in 2020.

Similar species: Could easily be mistaken for *Ocyptamus fascipennis* (which has complete bands on wings) or *Baccha cognata* (which has clear wings).

Flowers: Sneezeweed in Minnesota.

Flight Dates: August

Status: Rare

Genus *Scaeva* :: Smoothwings

The single Nearctic species of *Scaeva* is a black-and-white flower fly with swollen frons.

White-bowed Smoothwing

The White-bowed Smoothwing (***Scaeva affinis***) is a medium-sized, black-and-white, Holarctic flower fly. The oval abdomen has paired white markings that are curved like parentheses and which do not reach the abdomen margin. The coloration of this species sets it apart from nearly all other species in the Syrphinae which have yellow markings. The R4+5 vein is not straight but has a noticeable, upward curve. The larvae are entomophagous, consuming mostly aphids. One study reported a single larva consuming close to 500 aphids.

Field Notes: This species is thought to be a migrant in Europe. Also, rarely encountered, there exists a melanistic form, without the characteristic white spots on the abdomen, has been observed.

Similar species: *Scaeva selenitica* which has yellowish markings instead of white and *Eupeodes volucris*, a western, montane species, which differs in having a straighter R4+5 vein.

Flowers: Tends to prefer blue flowers. Has been observed on Blue Flax (*Linum lewisii*) in Minnesota.

Flight Dates: June

Status: Common

Range: Northern

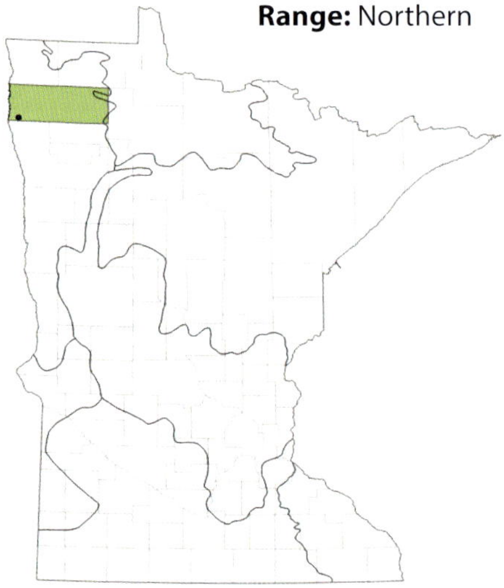

Genus *Sphaerophoria* :: Globetails

Small yellow-and-black syrphines with long, cylindrically-shaped abdomens. These distinct flies have a bright-yellow scutellum and yellow face. Males have large, globe-shaped genitalia.

Tufted Globetail

The Tufted Globetail (***Sphaerophoria contigua***) is a small, Nearctic species. The yellow stripes on the sides of the thorax end at the transverse suture, just above the wing base. Conspicuous tuft of hair on the rounded, dorsal lobe of male surstylus (*see below*).

Illustration of male surstylus

Field Notes: This species is sexually dimorphic. The males have abdomens that are thin, with the rear bands often blurred into solid yellow or a kind of fiery-orange. The females, on the other hand, have slightly broader abdomens with thin, arced bands.

Similar species: Most similar to other *Sphaerophoria* species.

Flowers: This wide-ranging species is found at a variety of habitats and flowers. In Minnesota, it has been recorded visiting flax, dandelion, sweet william, nodding wild onion, goldenrod, sedum, aster, tansy, downy gentian, ox-eye daisy, Canada honewort, and hawksbeard.

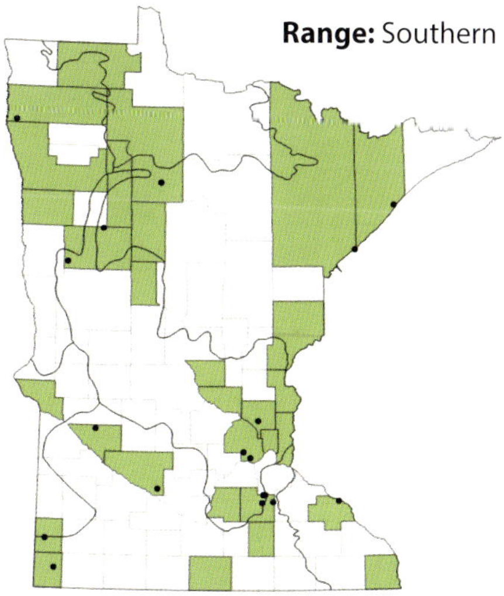

Range: Southern

Flight Dates: June - October. Most abundant in September.

Status: Abundant

Black-footed Globetail

The Black-footed Globetail (***Sphaerophoria philanthus***) is a small, Nearctic species. The yellow stripes on the sides of the thorax continue past the transverse suture, sometimes rather faintly. Less of a tuft of hair on the dorsal lobe of male surstylus and with a flattened process on the ventral lobe (*see drawing*). The apical tarsomeres of the front and middle feet are dark brown or black, sometimes all tarsomeres are darkened. Males often have a pair of distinct, diamond-shaped marks on the fifth abdominal section.

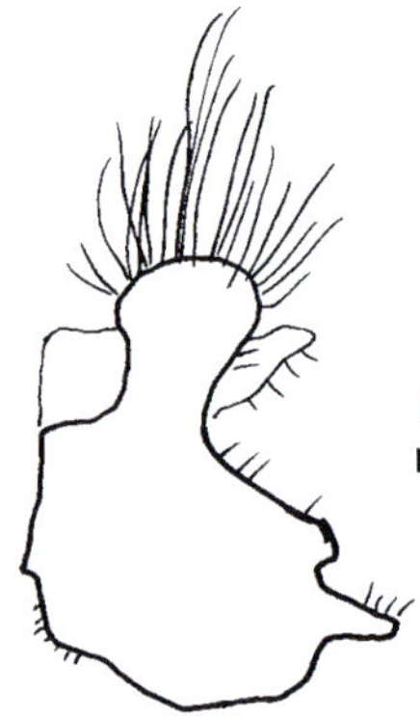

Illustration of male surstylus

Field Notes: Care must be taken in identifying all *Sphaerophoria* species. To be absolutely sure, specimens should be examined.

Similar species: Most similar to other *Sphaerophoria* species.

Flowers: Minnesota records include fleabane, wild plum, and goldenrod.

Flight Dates: May - August

Status: Uncommon

Range: Northern

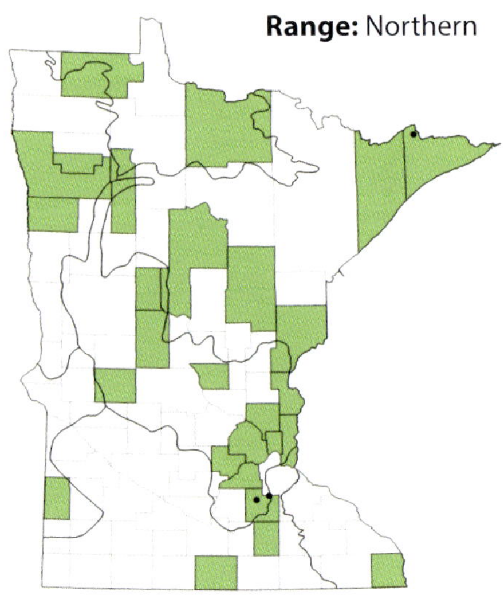

Genus *Syrphus* :: Flower Flies

The genus Syrphus includes species that have, possibly, the most quintessential appearance of flower flies: medium-sized, bright, yellow-and-black flies, conspicuous on garden flowers and at forest margins.

The character unique to this genus is the long hairs on the upper surface of the lower/rear calypter membrane. However, this character is difficult to see—the hairs are similar in color to the calypter surface so even under magnification the lighting needs to be just right to highlight the hairs. Once observed it is definitive. A more general character is the yellow "golf club" markings of tergite 2 and the yellow "mustache-like" markings of tergites 3 and 4. Differentiation of species relies upon microscopic examination of eyes, wing-cell trichosity, leg coloration and the color pattern of the abdomen margin.

The male often hovers one to several meters above the ground under trees or next to hills. The larvae feed on a variety of aphids and closely related insects.

Five species recorded in Minnesota with three additional species possible.

Key to Syrphus species

1 Eyes pilose .... ***Syrphus torvus***
1′ Eyes bare 2

2 Wing cell bm bare anteriorly 3
2′ Wing cell bm densely microtrichose 6

3 Yellow abdominal bands do not meet in the middle .... ***Syrphus sexmaculatus***
3′ Yellow abdominal bands on tergites 3 and 4 complete 4

4 Yellow abdominal bands do not meet margin .... ***Syrphus opinator***
4′ Yellow abdominal bands meet margin narrowly .... ***Syrphus vitripennis***
4″ Yellow abdominal bands meet margin broadly 5

5 Metafemur with pre-apical dark band .... ***Syrphus knabi***
5′ Metafemur (of male) completely dark on the basal two-thirds; Metafemur (of female) mostly yellow .... ***Syrphus rectus***

6 Yellow abdominal bands on tergites 3 and 4 complete .... ***Syrphus ribesii***
6′ Yellow abdominal bands do not meet in the middle .... ***Syrphus attenuatus***

Common Flower Fly

The Common Flower Fly (***Syrphus ribesii***) is a medium-sized, Holarctic flower fly. The eyes are bare. The wing cell, bm, is densely and completely microtrichose. Complete yellow bands on the third and fourth abdominal segments and an alternating yellow and black abdominal margin. Males, in Minnesota, often have partial, black medial line on face. The larvae are entomophagous, consuming aphids. Males, in Minnesota, often have partial, black medial line on face.

Field Notes: Polyvoltine species with larvae that exhibit facultative diapause. Researchers have shown overwintering larvae to survive temperatures as low as -35° C (-31° F).

Similar species: Most similar to other yellow-and-black syrphines, especially *Syrphus, Eupeodes,* and *Epistrophe*. Positive identification can only be made with careful microscopic examination.

Flowers: Adults of genus Syrphus have been observed at a wide variety of flowers, including, in Minnesota, white snakeroot, dandelion, goldenrod, aster, ox-eye daisy, purple coneflower, tansy, and wild onion.

Flight Dates: May - October

Range: Northern

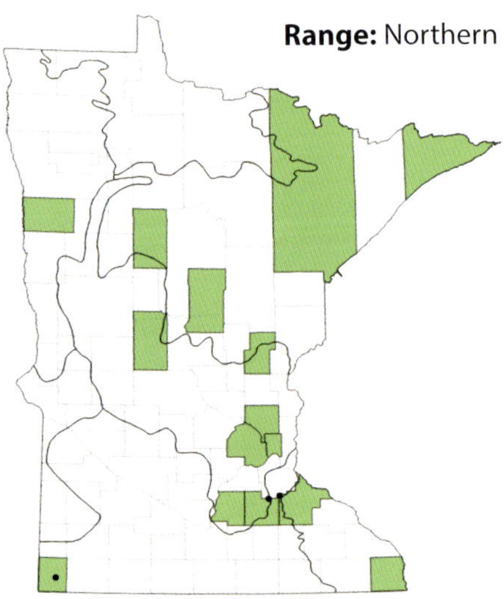

Status: Common

Yellow-legged Flower Fly

The Yellow-legged Flower Fly (***Syrphus rectus***) is a medium-sized, Nearctic species. The eyes are bare. The wing cell, bm, is mostly bare. Complete yellow bands on the third and fourth abdominal segments form an alternating yellow and black abdominal margin. The female hind femur is yellow in most individuals; the male hind femur is dark, with yellow only on the apical end. Larvae feed on aphids.

Field Notes: Adults are found in a wide variety of habitats from bogs to urban forests.

Similar species: Most similar to *Syrphus vitripennis* but can easily be confused with other yellow and black syrphines. Microscopic examination usually needed to make a positive identification.

Flowers: Adults have been observed on mock orange in Minnesota. Known to visit asters elsewhere.

Flight Dates: June - August

Range: Eastern

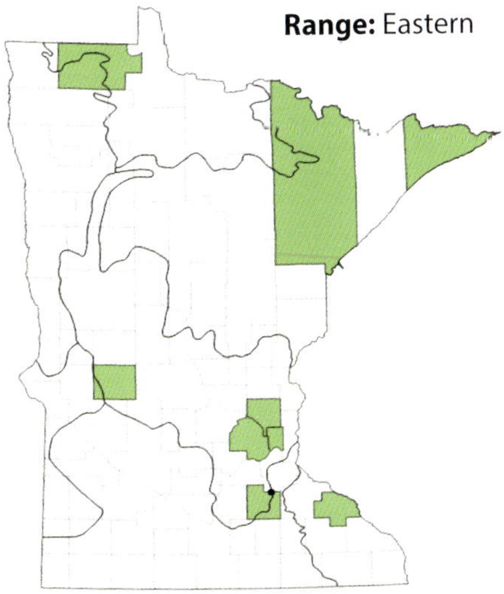

Status: Common

Six-spotted Flower Fly

The Six-spotted Flower Fly (***Syrphus sexmaculatus***) is a medium-sized, Holarctic flower fly. The eyes are bare. The wing cell, bm, is mostly bare. The yellow bands on the second, third and fourth abdominal segments are divided medially. The yellow bands are narrow where they cross the abdominal margin. Larvae feed on aphids.

Field Notes: Found in the mixed boreal region of northern Minnesota. Rare.

Similar species: Most similar to *S. attenuatus* but also other yellow-and-black syrphines, especially *Syrphus, Eupeodes*, and *Epistrophe*. Positive identification can only be made with careful microscopic examination.

Flowers: Adults, elsewhere, have been observed visiting flowers of the family Apiaceae.

Flight Dates: September

Range: Northern

Status: Rare

Additional Syrphus Minnesota Species Not Included in the Guide

Hairy-eyed Flower Fly **(*Syrphus torvus*):** Northern species; pilose eyes

Syrphus attenuatus: * Western species; similar to *S. sexmaculatus*

Black-legged Flower Fly **(*Syrphus vitripennis*):** Northern species

Black-margined Flower Fly ***Syrphus opinator:*** Western species

Eastern Flower Fly ***Syrphus knabi:*** Eastern species

Black-margined Flower Fly

Genus *Xanthogramma* :: Harlequins

The flies of the genus *Xanthogramma* are black-and-yellow wasp mimics. They have yellow faces and a yellow stripe on the side of the thorax. They are usually found close to the ground near the *Lasius* ant colonies where the larvae feed on aphids tended by the ants.

American Harlequin

The American Harlequin (***Xanthogramma flavipes***) is a medium-sized, brightly-colored, Nearctic species. It has bright-yellow stripes on the sides of the thorax. The scutellum has a yellow margin. The abdomen is marked by thin, complete bands, making it a convincing potter wasp mimic. Note that in some individuals these thin yellow bands are paired with an additional dark orange or brown band. The yellow colors of this species are more lemony-yellow than other flower flies.

Field Notes: This fascinating fly is often found near the ground in low vegetation where it can sometimes be observed exploring the entrances of ant hills where it presumably oviposits.

Similar species: While similar to other black-and-yellow syrphines, it most resembles the very rare, and closely-related species *Doros aequalis*.

Flowers: Adults rarely observed visiting flowers.

Flight Dates: May - August

Range: Eastern

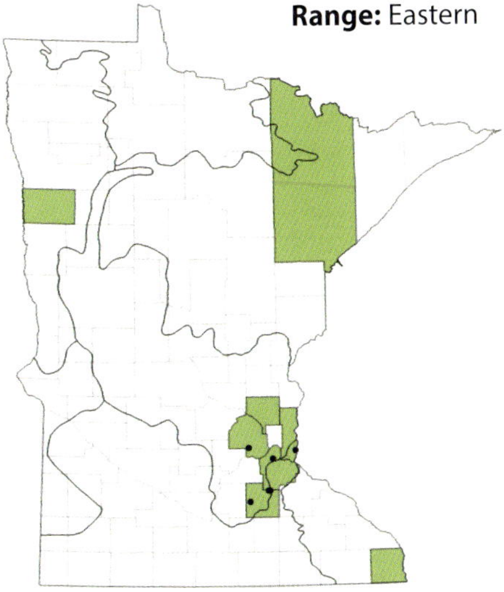

Status: Uncommon

Tribe Toxomerini

Tribe Toxomerini includes two genera in Minnesota: *Allograpta* and *Toxomerus*. The four Minnesota species in this tribe are some of the most abundant and most commonly encountered flower flies.

Genus *Allograpta* :: Streaktails

The flies of the genus *Allograpta* are small, black-and-yellow flies. A single species is known from Minnesota and is easily recognizable by the distinct abdominal markings.

Oblique Streaktail

The Oblique Streaktail (***Allograpta obliqua***) is a small Nearctic flower fly. Abdomen section 4 has two medial yellow stripes and two oblique stripes. This pattern is unique.

Field Notes: Often perches conspicuously on blooming composites.

Similar species: Most similar to other small yellow and black species such as *Toxomerus* and *Sphaerophoria*.

Status: Common

Flowers: Adults have be observed at a wide variety of flowers in Minnesota.

Flight Dates: May - October

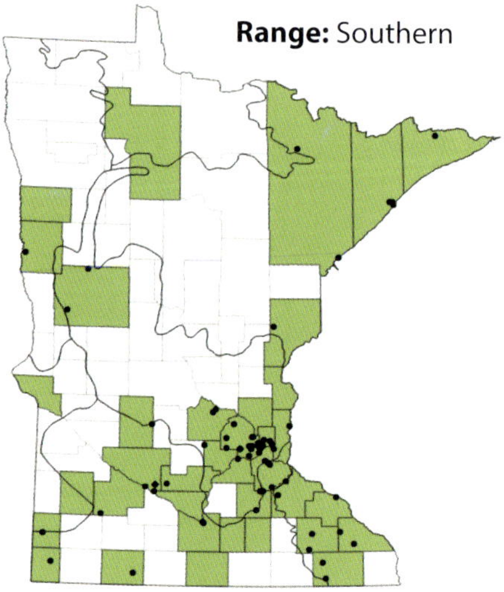

Range: Southern

Genus *Toxomerus* :: Calligraphers

A Nearctic genus with the highest diversity in species occurring in the South American tropics. Three species are known from Minnesota.

Margined Calligrapher

The Margined Calligrapher (***Toxomerus marginatus***) is a small, Nearctic flower fly. It is recognized by the solid, yellow margin around the abdomen. This margin is sometimes difficult to see. The pattern of the abdomen is unique. The marks on segments 2 and 3 are usually full and rounded and separated by a yellow median stripe.

Field Notes: Ubiquitous on flowers of low vegetation. The most commonly encountered flower fly in Minnesota.

Similar species: Most similar to Eastern Calligrapher.

Flowers: Adults are very often encountered on dandelions, but visit a wide variety of other flowering plants.

Flight Dates: April - October

Status: Abundant

Range: Southern

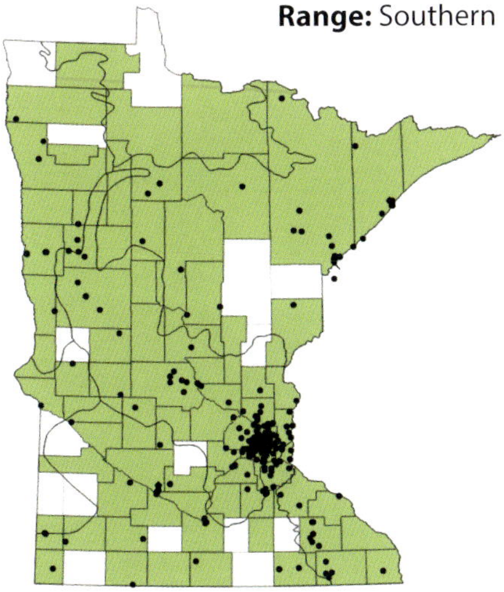

Eastern Calligrapher

The Eastern Calligrapher (***Toxomerus geminatus***) is a small, Nearctic flower fly. It is recognized by its abdomen margin which alternates yellow and black. The yellow markings on segments 3 and 4 are somewhat abbreviated and slightly more angular than the pattern found on *S. marginatus*.

Field Notes: Males and females often observed in tandem, both perched and in flight.

Similar species: Most similar to Margined Calligrapher.

Flowers: Adults are very often encountered on dandelions, but visit a wide variety of other flowering plants.

Flight Dates: April - October

Status: Abundant

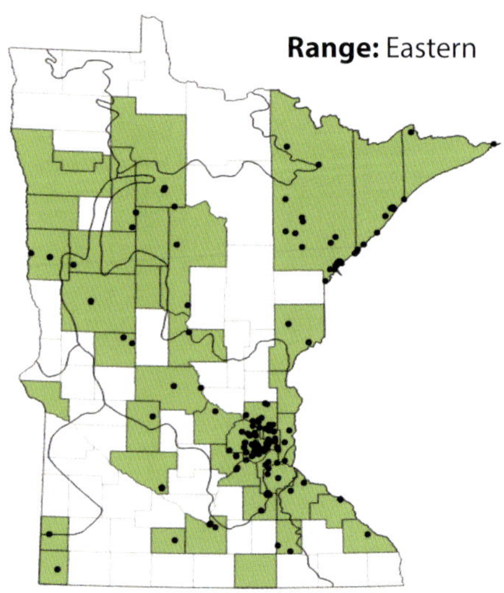

Range: Eastern

Maize Calligrapher

The Maize Calligrapher (***Toxomerus politus***) is a small, Nearctic flower fly. It seems to reach its northern limit in Minnesota where most records are limited to the southern half of the state. There is a pronounced pollinose medial stripe on the thorax. Abdominal segments 2-4 have a pair of bands, one thin and one thick, the thick band is divided by a medial yellow mark.

Field Notes: The larvae of this species are known to be phytophagous and not carnivorous, feeding on pollen in the seedhead of corn. Given its association with crops, it wouldn't be surprising if this species becomes much more rare with the application of modern agricultural chemicals.

Similar species: Shouldn't be mistaken for any other flower fly as the pattern on the abdomen is unique.

Flowers: In Minnesota, adults have been observed on grass blades, at ground ivy and ornamental allium flowers.

Flight Dates: August - September

Status: Uncommon

Range: Southern

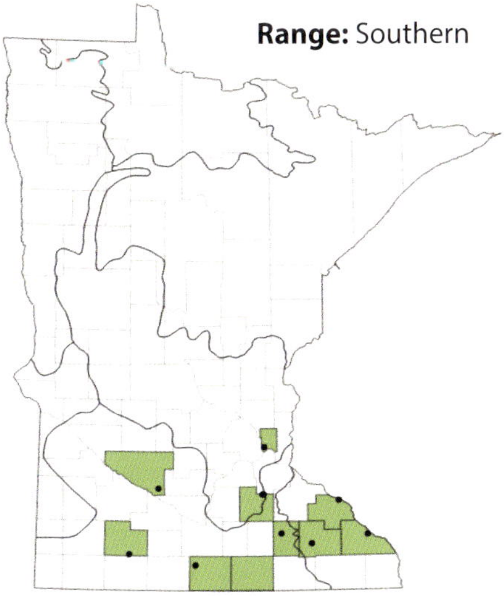

Chapter Three
PIPIZINAE

PIPIZINAE

Pipizinae Genera Summary

Tribe	Genera
Pipizini	*Heringia, Neocnemodon, Pipiza, Trichopsomyia*

The subfamily Pipizinae includes some of our most interesting and elusive flower flies. Historically, this group of flies has been placed as a tribe in subfamily Eristalinae for morphological reasons or, alternatively, in subfamily Syrphinae because of similarities in larvae behavior. However, recent phylogenetic research shows that the flies of tribe Pipizini make up a monophyletic sister group to the subfamily Syrphinae and should be given its own subfamily status (Mengual et al., 2015).

The pipizines are small- to medium-sized, black flies. The face is flat, without any protrusion. They have a rounded oral margin and small, facial pits near the bottom edge of the eyes. They tend to bask with their wings spread and they are often photographed in this delta-winged position. The larvae have specialized to prey on gall-forming aphids and psyllids, often feeding inside the galls and leaf-curls.

The Pipizinae are represented by four genera in Minnesota: *Heringia, Neocnemodon, Pipiza*, and *Trichopsomyia*. Most pipizines have an early flight season, with most observations occurring in May.

Genus *Heringia* :: Smoothlegs

The flies of the genus *Heringia* are small, entirely black flies. The larvae are entomophagous, thought to prey upon gall-forming or leaf-rolling aphids and psyllids, typically on poplar or elm. These flies lack the spur-like projections on the trochanters that are found on *Neocnemodon*.

Eastern Smoothleg

The Eastern Smoothleg (***Heringia salax***) is a small, black Nearctic flower fly. The flagellum is longer than wide. Brownish, yellowish pile on the thorax. The male hind femur is uniform in width and does not have sharp, ventral-apical projection. Larvae are entomophagous, preying on gall-forming aphids.

Field Notes: Larvae are known predators of wooly apple aphids and grape phylloxera.

Similar species: Difficult to separate from other species of *Neocnemodon* and *Heringia* without close, Microscopic scrutiny of specimens.

Flowers: No Minnesota floral records.

Flight Dates: Unknown

Status: Uncommon

Range: Eastern

Additional Minnesota Heringia Species Not Included in the Guide

Heringia canadensis: * Eastern species

Genus *Neocnemodon* :: Spikelegs

The flies of the genus *Neocnemodon* are small, entirely black flies. The larvae are entomophagous, thought to prey upon coccids, aphids, and adelgid plant bugs, typically on conifer trees. These flies have spur-like projections on the trochanters of the hind legs and sometimes the middle legs as well. Sometimes treated taxonomically as a subgenus of *Heringia*.

Elongate Spikeleg

The Elongate Spikeleg (***Neocnemodon elongata***) is a small, black Nearctic flower fly. Thin projection on the hind trochanters, the third antennae segment roundish, and the mid-tibia only slightly expanded.

Field Notes: While very few records exist, these flies are most likely quite common and simply overlooked. Search for them in small, sunlit clearings in woods where they are sometimes found visiting moist ground.

Similar species: Difficult to separate from other species of *Neocnemodon* and *Heringia* without close, Microscopic scrutiny of specimens.

Flowers: *Neocnemodon* sp. has been observed on bush honeysuckle in northern Minnesota.

Flight Dates: May

Range: Northern

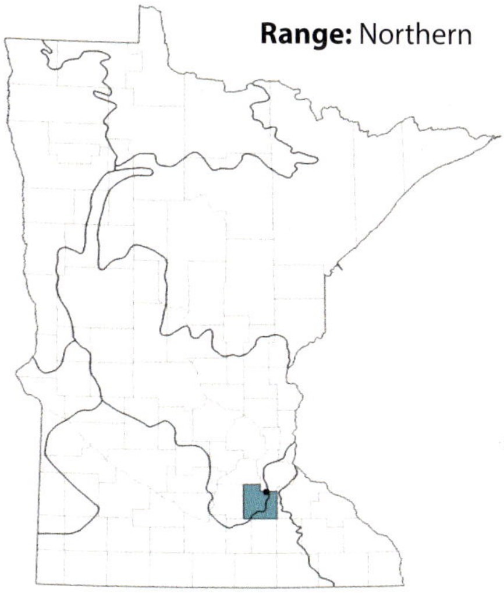

Status: Uncommon

Additional Minnesota Neocnemodon Species Not Included in the Guide

Neocnemodon calcarata, Opaque Spikeleg; ***Neocnemodon carinata***, Carinate Spikeleg; ***Neocnemodon coxalis***, White-faced Spikeleg; ***Neocnemodon intermedia***, Midnight Spikeleg; ***Neocnemodon pisticoides***, Long-spined Spikeleg; ***Neocnemodon rita***, Black-faced Spikeleg *

Genus *Pipiza* :: Pitheads

The flies of the genus *Pipiza* are medium-sized, black flies. The frons is slightly projected, creating an angular profile to the face. Wings occasionally have dark or iridescent-blue clouds. Abdomens are shiny black. Many species have a pair of spots on abdomen segment 2, but these are often quite variable making identification difficult. The larvae are entomophagous, thought to prey upon gall-forming aphids, typically on poplar or elm.

Four-spotted Pithead

The Four-spotted Pithead (***Pipiza quadrimaculata***) is a small, Holarctic flower fly. The abdomen is broadly oval and usually has two pairs of yellow spots. Antennae are attached at or below the midpoint of the head. The hind femur is only slightly swollen.

Field Notes: Prefers shadier habitats than other *Pipiza* species.

Similar species: Very similar to *Pipiza femoralis* and other larger pipizines. Also could be confused with flies of the genus *Blera* but differs in wing venation.

Range: Northern

Flowers: No Minnesota floral records. Elsewhere, the adults visit a variety of flowers.

Flight Dates: Unknown

Status: Hypothetical

White-haired Pithead

The White-haired Pithead (***Pipiza femoralis***) is a medium-sized, Nearctic flower fly. Abdomen is black with a pair of yellowish-green spots on segment two. The male abdomen is often entirely black or with a smaller, less-visible pair of spots. The hind femur widens apically with a ridge on the underside which ends in a visibly-sharp angle. There are no black hairs on the femur (as with *P. puella*). The male thorax is covered in white pile; the female pile varies in color from white to golden yellow.

Field Notes: Not frequently observed but can be locally abundant. Probably the most common pipizine in Minnesota.

Similar species: Very similar to *Pipiza puella* and other larger pipizines. It could also be confused with flies of the genus *Blera* but differs in wing venation.

Flowers: Only Minnesota floral record is dandelion. Elsewhere, flowers visited include garlic mustard, hawthorn, plum, and currant.

Flight Dates: May - June

Range: Eastern

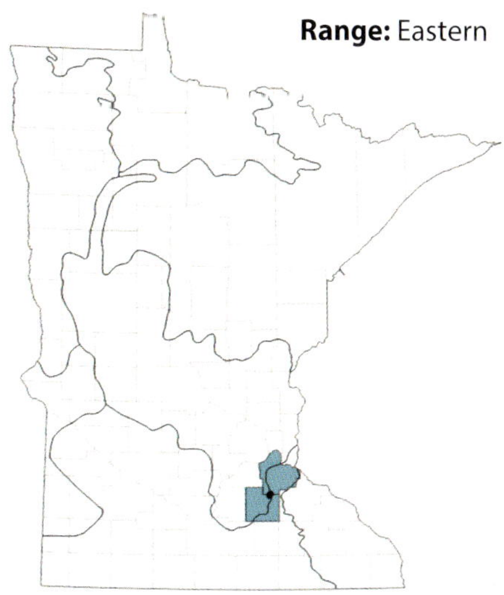

Status: Abundant

Sumac Gall Pithead

The Sumac Gall Pithead (***Pipiza puella***) is a medium-sized, Nearctic flower fly. The male abdomen is black. The female abdomen is black and usually has a pair of small yellow spots on segment two. The hind femur widens apically with a ridge on the underside which ends in a visibly-sharp angle and with black pile, also apically. Larvae are associated with aphids causing sumac galls.

Field Notes: The larvae are presumed to be specialists that prey upon staghorn sumac aphids (*Melaphis rhois*), a variety of gall-forming, wooly aphid.

Similar species: Very similar to *Pipiza femoralis* and other larger pipizines. It could also be confused with flies of the genus *Blera* but differs in wing venation.

Flowers: Wild geranium and prairie phlox in Minnesota. Elsewhere, the adults visit a variety of flowers.

Flight Dates: June

Range: Northern

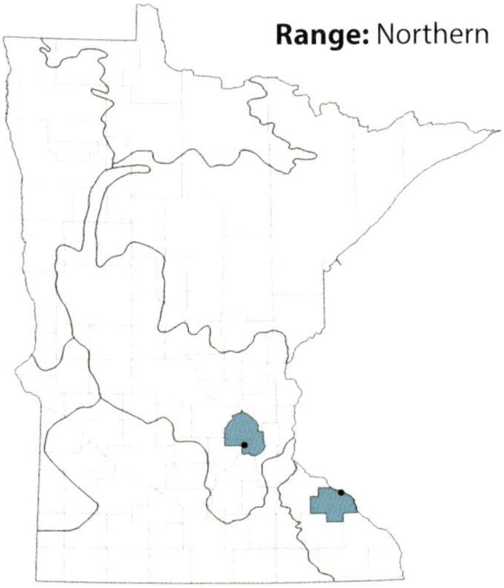

Status: Rare

Additional Minnesota Pipiza Species Not Included in the Guide

Pipiza atrata: Northern species; Female abdomen always black

Pipiza macrofemoralis: *Northern species

Genus *Trichopsomyia* :: Psyllid-killers

The flies of the genus *Trichopsomyia* are small, entirely black flies. They have long hairs on the anterior anepisternum just behind the anterior respiratory spiracle. Wing venation differs from the other three genera in that the M1 crossvein meets R4+5 at close to 90 degrees (this angle is closer to 45 degrees in the other genera). The larvae are entomophagous, preying upon gall-forming psyllids and aphids.

Black-haired Psyllid-Killer

Black-haired Psyllid-Killer (***Trichopsomyia apisaon***) is a small, black-bodied, Nearctic flower fly. The anterior anepisternum has long pile—the defining character of this genus. The black pile on the frontal triangle points to this species. Definitely not easy to identify, even with a microscope. The larvae prey upon gall-forming aphids and psyllids.

Range: Eastern

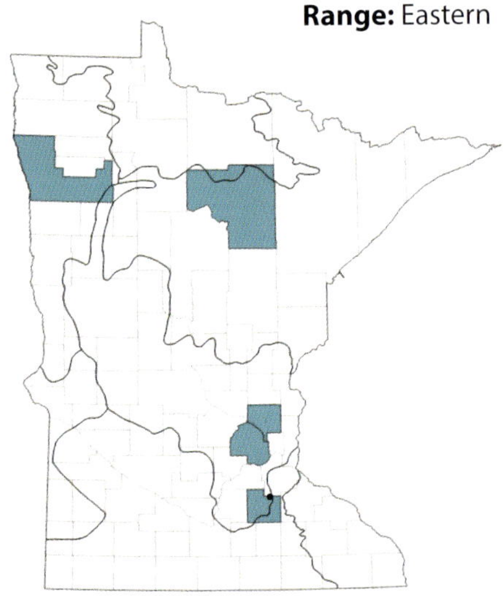

Field Notes: Adults do visit flowers but are most often encountered basking on sunlit leaves near the trees that host the psyllids that the larvae prey upon.

Similar species: Very similar to other Pipizines, must be separated by microscopic characters.

Flowers: No Minnesota floral records.

Flight Dates: June

Status: Uncommon

Additional Minnesota Trichopsomyia Species Not Included in the Guide

Trichopsomyia banksi: * Southern species; eyes touching, white pile on face

Trichopsomyia pubescens: * Southern species; eyes narrowly separated, white pile on face

Trichopsomyia recedens: Southern species; eyes touching, white pile on face

Chapter Four
ERISTALINAE

ERISTALINAE

Eristalinae Genera Summary

Tribe	Subtribe	Genera	Pages
Brachyopini	**Brachyopina**	*Brachyopa, Chrysogaster, Chrysosyrphus, Hammerschmidtia, Myolepta, Orthonevra*	118-127
	Spheginina	*Neoascia, Sphegina*	128-131
Cerioidini		*Ceriana, Sphiximorpha*	132-133
Eristalini	**Eristalina**	*Eristalinus, Eristalis*	134-147
	Helophilina	*Arctosyrphus, Anasimyia, Eurimyia, Helophilus, Mallota, Parhelophilus, Polydontomyia*	148-160
	Sericomyiina	*Sericomyia*	161-164
Merodontini		*Eumerus, Merodon*	165-167
Milesiini	**Blerina**	*Somula, Blera*	168-171
	Criorhinina	*Criorhina, Sphecomyia*	172-173
	Milesiina	*Milesia, Spilomyia*	174-178
	Temnostomina	*Pterallastes, Temnostoma, Teuchocnemis*	179-186
	Tropidiina	*Syritta, Tropidia*	187-188
	Xylotina	*Brachypalpus, Chalcosyrphus, Xylota*	189-203
Rhingiini	**Cheilosiina**	*Cheilosia, Ferdinandea*	204-208
	Rhingiina	*Rhingia*	209-210
Volucellini		*Copestylum, Volucella*	211-214

Key to the genera of the subfamily Eristalinae

1 Antennae are long 2
1′ Antennae are short 5

2 Arista projects from the end of the antennae 3
2′ Arista projects from the dorsal surface of the flagellum .... ***Sphecomyia***

3 Eyes pilose. Abdomen a solid, brassy color .... ***Callicera***
3′ Eyes bare. Abdomen is black with yellow bands 4

4 Antennae are long, attached to short prominence .... ***Sphiximorpha***
4′ Antennae are short but attached to a long prominence .... ***Ceriana***

5 The underside of hind femur, apically, has a large, flat, angular flange 6
5′ No flat, angular flange on the hind femur, though spines or smaller, round protuberances may be present 7

6 Eyes bare. Face with a central ridge .... ***Tropidia***
6′ Eyes pilose. Face flat, without central ridge. Crossvein M2 angles back where it meets R4+5 .... ***Merodon***

7 Wing vein R4+5 deeply sinuous 8
7′ Wing vein R4+5 is straight 17

8 Wing cell r1 is closed, connected to wing margin by short vein 9
8′ Wing cell r1 is open or closed at the wing margin 10

9 Scutellum is black. Eyes have numerous black spots .... ***Eristalinus***
9′ Scutellum is yellow or brown. Eyes without spots .... ***Eristalis***

10 Thorax uniformly colored 11
10′ Thorax with pollinose, longitudinal stripes on dorsum 12

11 Hind femurs inflated .... ***Mallota***
11′ Hind femurs straight, not inflated .... ***Pterallastes***

12 Wide, diffuse pterostigma. Wing cell r1 opening to wing margin is narrow .... ***Helophilus***
12′ Narrow, distinct pterostigma. Wing cell r1 opening to wing margin is wide ... 13

Key to the genera of the subfamily Eristalinae continued

13 Rear femur enlarged and strongly curved .... ***Polydontomyia***
13′ Rear femur can be enlarged but not strongly curved 14

14 Face greatly produced 15
14′ Face not greatly produced 16

15 Face produced forward, extending beyond antennae .... ***Eurimyia***
15′ Face produced downward .... ***Arctosyrphus***

16 Hind tibia yellow, usually with two dark rings .... ***Anasimyia***
16′ Hind tibia yellow, yellow with keel on basal half, usually with one dark ring .... ***Parhelophilus***

17 Arista is plumose 18
17′ Arista is bare or with short hairs 19

18 Wing cell r1 closed .... ***Volucella***
18′ Wing cell r1 open to wing margin .... ***Sericomyia***

19 Anterior crossvein r-m joins cell dm distally or at midpoint 20
19′ Anterior crossvein r-m joins cell dm well before midpoint 28

20 Face without central knob and without produced mouth .... ***Eumerus***
20′ Lower part of face produced outward or downward or both 21

21 Lower part of face distinctly produced forward. Mouth margin strongly produced and turned downward .... ***Criorhina***
21′ Mouth margin not extended downwards. With or without facial knob 22

22 Face more or less entirely yellow. Tufts of bristles on thorax and scutellum .... ***Ferdinandea***
22′ Face black or yellow with black medial stripe... 23

23 Thorax with distinct yellow marks—dots, triangles, dashes. Social wasp mimic 24
23′ Thorax without yellow marks 25

24 Underside of hind femur with a spine near apex. Patterned eyes .... ***Spilomyia***
24′ Hind femur without spine. Eyes without patterning .... ***Temnostoma***

Key to the genera of the subfamily Eristalinae continued

25 Abdomen is widely oval-shaped .... ***Blera***
25′ Abdomen is elongate, parallel-sided 26

26 Hind femur slightly swollen. Metasternum with short pile .... ***Xylota***
26′ Hind femur distinctly swollen. Metasternum with long hairs 27

27 Small species. Face with central ridge .... ***Syritta***
27′ Small or large species. Head wider than deep.
Face without central ridge .... ***Chalcosyrphus***

28 Abdomen is slender, anterior part constricted 29
28′ Abdomen oval- or parallel-sided 30

29 Angle between outer crossveins M1 and vein R4+5 is orthogonal .... ***Neoascia***
29′ Angle between outer crossveins M1 and vein R4+5 is acute .... ***Sphegina***

30 Lower part of face strongly produced, beak-like .... ***Rhingia***
30′ Lower part of face not beak-like 31

31 Face with a distinct zygoma (eye margin) and distinct facial knob .... ***Cheilosia***
31′ Face without zygoma 32

32 Abdomen or scutellum is yellow, orange, or yellowish-brown 33
32′ Abdomen is dark, black, or metallic green. Scutellum is black 34

33 Outer crossvein M1 is rounded. Large species .... ***Hammerschmidtia***
33′ Outer crossvein M1 is angular. Small or medium species .... ***Brachyopa***

34 Upper crossvein M1 angles anteriorly before it meets R4+5 .... ***Orthonevra***
34′ Upper crossvein M1 curves opposite direction where it meets R4+5 35

35 Antennae attached at the upper third of the head in profile .... ***Chrysosyrphus***
35′ Antennae attached at the middle or lower part of head
in profile .... ***Chrysogaster***

Tribe Brachyopini

Tribe Brachyopini includes eight genera in Minnesota, further separated into two subtribes: subtribe Brachyopina (*Brachyopa, Chrysogaster, Chrysosyrphus, Hammerschmidtia, Myolepta, Orthonevra*); and Subtribe Spheginina (*Neoascia, Sphegina*).

The middle of the face is strongly concave without a central knob (as in *Cheilosia*).

Genus *Brachyopa* :: Sapeaters

The flies of the genus *Brachyopa* are small, oddly-shaped flies that are also unusually colored (compared to other flower flies)—pale-yellow, orange, or grayish-brown. The lower part of the face protrudes forward and the eyes are large and shaped rather like goggles. The third antennal segment has a pit with sensory cells. The arista is bare for some species and slightly plumose for other species. The wings are relatively long, extending well beyond the end of the abdomen.

A number of factors combine to make these flies difficult to find: short flight period for adults, associated with sap-runs of wounded trees (instead of routinely visiting flowers), adults are small, inconspicuous, swift fliers that resemble, because of their color and shape, flies of families other than Syrphidae, such as Scathophagidae and Dryomyzidea for example. The majority of observations are made at sap runs where the adult flies hover in front of the wound, perch on the dry bark adjacent the sap run, or oviposit directly in the flowing sap.

The larvae live in sap runs of living trees and collections of sap under the bark of fallen trees. The larvae are somewhat flattened and have a relatively long posterior breathing tube. The final few segments are fringed with lappets.

Count yourself lucky if you encounter one of these rare and stealthy flower flies!

Yellow-spotted Sapeater

The Yellow-spotted Sapeater (***Brachyopa vacua***) is a small, brown Nearctic species. The thorax is gray with purplish overtones. Segment two of the abdomen is almost entirely yellow, while segment three has small spots of yellow on the anterior corners. Goggle-eyed, slightly irregular in shape, the eyes sweep forward along the produced face. Most specimens have been found at elm tree wounds. The flies are observed hovering in front of the wound, perched adjacent to the flow of sap, or ovipositing directly into the flowing sap. The larvae feed on organisms in the fermenting sap and live in the wounds.

Field Notes: The bi-colored abdomen visible through the folded wings, at first glance, creates the illusion that the wings are banded.

Similar species: Most similar to other *Brachyopa* species. Also could be confused with *Hammerschmidtia rufa*.

Flowers: Adults are most commonly observed at flowing wounds on elm trees, but they have occasionally been observed at flowers, including Goats Beard, Dogwood, and Cow Parsnip.

Flight Dates: June

Range: Southern

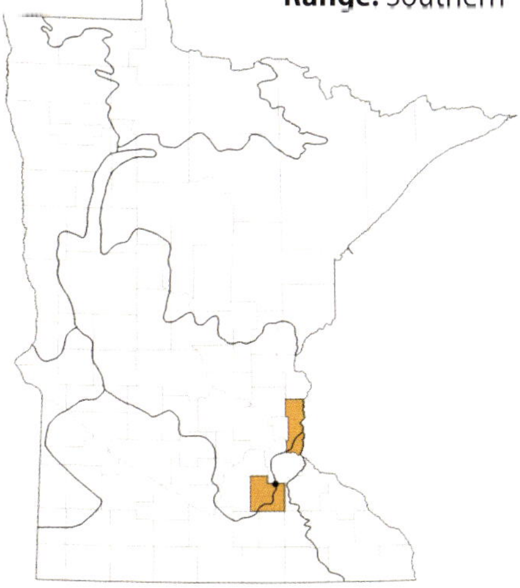

Status: Rare

Plain-winged Sapeater

The Plain-winged Sapeater (***Brachyopa caesariata***) is a small, brown Nearctic species. Only recently described in 2020, this species has a mostly-yellow appearance. The arista is pubescent. The katepisternum is pilose on the upper half. The abdomen is mostly yellow with faint black markings or solid thin black markings in a complex pattern. Goggle-eyed, slightly irregular in shape, the eyes sweep forward along the produced face. The larvae feed on the sap and live in the wounds of deciduous trees, possibly alder and poplar.

Field Notes: The two photo observations for this species in Minnesota show it perched on leaves of chokecherry and fern probably near its host tree.

Similar species: Most similar to other *Brachyopa* species. Also could be confused with *Hammerschmidtia rufa*.

Flowers: Adults will most likely be observed at flowing wounds on deciduous trees, only rarely have they been observed at flowers, Heracleum and Rubus elsewhere.

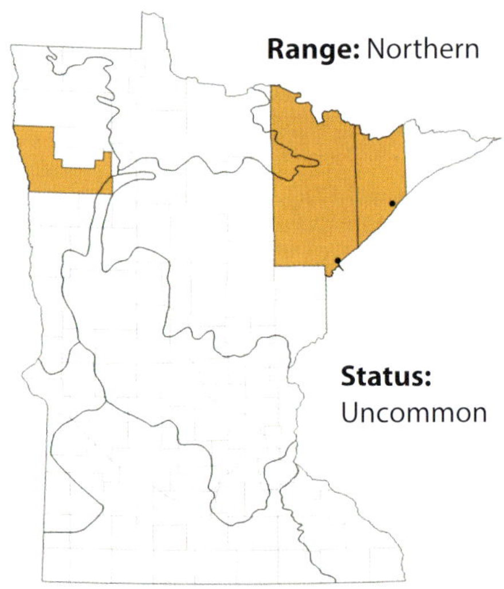

Range: Northern

Status: Uncommon

Flight Dates: June

Additional Minnesota Species Not Included in the Guide

Brachyopa notata: Northern species. Yellow abdomen with a medial stripe joined to a thin, dark apical bands on each segment.

Genus *Hammerschmidtia* :: Logsitters

The flies of the genus *Hammerschmidtia* are medium-sized, brownish-orange flower flies that resemble, but are not closely related to, the genus *Brachyopa*. They differ in being larger with a longer abdomen and by the presence of long bristles on the thorax in front of the wings and bristles on the scutellum. They inhabit mixed boreal forests and are associated with old growths of aspen, often perching on stumps and fallen logs.

Black-bristled Logsitter

The Black-bristled Logsitter (***Hammerschmidtia rufa***) is a medium-sized, Nearctic flower fly. The abdomen color is variable, ranging from orange to nearly black. The long bristles on the thorax are black. Similar bristles are also present on scutellum. Face and eye shape resemble *Brachyopa*.

Field Notes:
The closely related Palearctic species, *H. ferruginea*, is known (from studies in Scotland) to breed in two microhabitats: sap-runs in living aspen trees and areas of wet decay under the bark of dead or fallen aspen trees.

Range: Northern

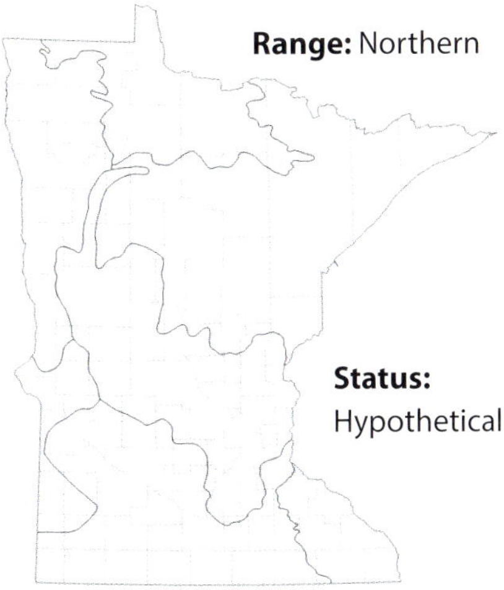

Status: Hypothetical

Similar species: Most similar to the undescribed Pale-bristled Logsitter, and can easily be confused with *Brachyopa* species.

Flowers: Adults are more frequently found at breeding sites than flowers but have been observed and collected at raspberry and wild plum.

Flight Dates: May - July

Genus *Chrysogaster* :: Wrinkleheads

The flies of the genus *Chrysogaster* are small, entirely-black flies with reddish-orange eyes. The flies are often shiny black with metallic sheens. The antennae are located at or below the midway point of the face. The female frons is heavily wrinkled. A single species is known from Minnesota. The larvae live underwater and have a short breathing tube.

Short-haired Wrinklehead

The Short-haired Wrinklehead (***Chrysogaster antitheus***) is a small, Nearctic flower fly. Male has short pile on the second sternite. Female has sparse, short pile on the first sternite. The female has a visibly wrinkled frons.

Field Notes: Habitat includes a range of wetlands and forest types.

Similar species: Very similar in size and shape to *Chrysosyrphus* and *Orthonevra*.

Flowers: Common ninebark, Canada honewort, and goutweed in Minnesota.

Flight Dates: May - June

Status: Common

Range: Eastern

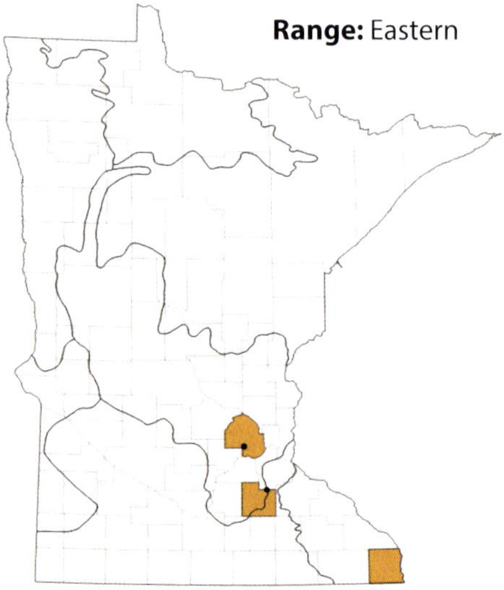

Genus *Chrysosyrphus* :: Wrinkleheads

The small, entirely-black flies of the genus *Chrysosyrphus* have only recently been separated from the genus *Chrysogaster*. This taxonomic split is supported by DNA evidence, but also the morphological traits of antennae placement and hairs on the scutellum. The antennae are located closer to the top of the head, significantly above the midpoint of the face. The scutellum has a row of small hairs under its posterior rim.

Variable Wrinklehead

The Variable Wrinklehead (***Chrysosyrphus latus***) is a small, Nearctic flower fly. The antennae are located closer to the top of the head, significantly above the midpoint of the face. The scutellum has a row of small hairs under its posterior rim. Pile on thorax and abdomen can vary in color from black to golden yellow. Male frons has a margin of pile.

Range: Northern

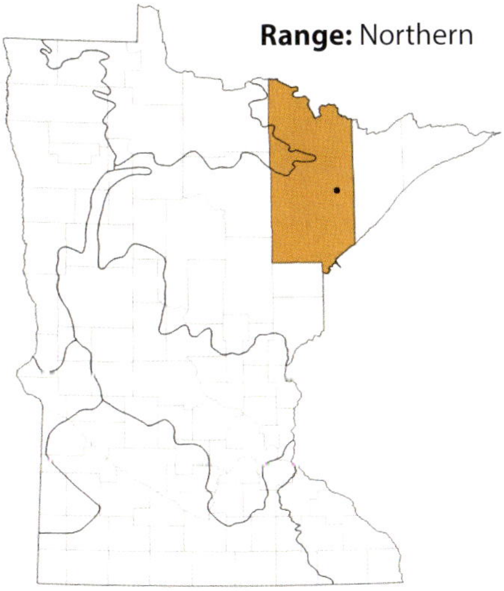

Field Notes: Bogs are the preferred habitat.

Similar species: Very similar in size and shape to *Chrysogaster* and *Orthonevra*.

Flowers: Willow catkins in Minnesota.

Flight Dates: May - June

Status: Rare

Genus *Orthonevra* :: Mucksuckers

The flies of the genus *Orthonevra* are very small, dark flower flies. The thorax and abdomen have a bronze metallic sheen. The eyes can be simply or complexly patterned. The larvae inhabit a number of high-organic, mucky situations.

Wavy Mucksucker

The Wavy Mucksucker (***Orthonevra nitida***) is a small, bronze-colored Nearctic flower fly. This fly is easily recognized by the elaborate pattern on its eyes—a simple horizontal line bisected by two wavy vertical lines. Wing crossveins are often darkened.

Field Notes: Very easy to overlook. Search for this remarkable fly near its breeding habitat at the shores of small, forested ponds with rich organic sediments.

Similar species: Because of its unique eyes it shouldn't be confused with similar species.

Flowers: Chrysanthemums in Minnesota.

Flight Dates: May - October

Status: Common

Range: Southern

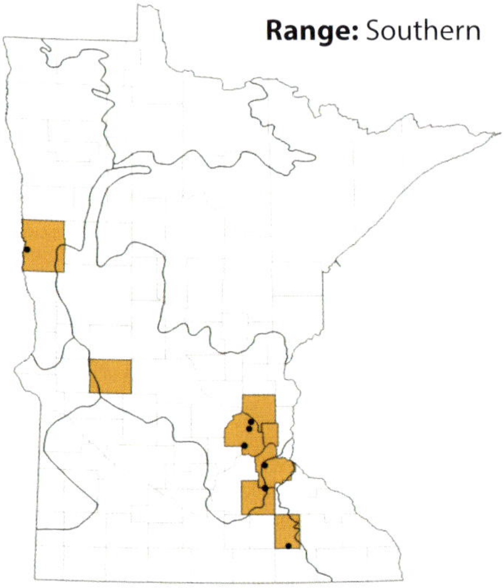

Beautiful Mucksucker

The Beautiful Mucksucker (***Orthonevra pulchella***) is a small, bronze-colored, Nearctic flower fly. The eyes have a single, horizontal line. Wings without darkening of crossveins. Very limited area of facial polinosity. Front and middle tibia are black in the middle, rear tibia yellow.

Field Notes: Associated with a variety of wetland and forest habitats.

Similar species: Most similar to *O. picti-pennis.*

Flowers: Goat's beard in Minnesota.

Flight Dates: June - July

Status: Uncommon

Range: Northern

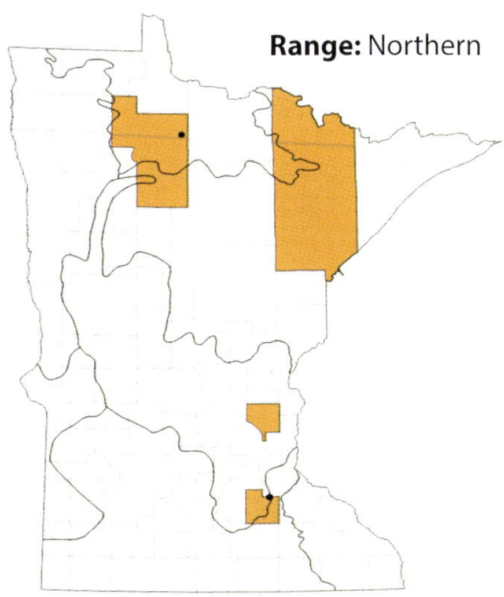

Dusky-veined Mucksucker

The Dusky-veined Mucksucker (***Orthonevra pictipennis***) is a small, bronze-colored, Nearctic flower fly. The eyes have a single, horizontal line. Wings crossveins are darkened. Wide area of facial polinosity. All tibia are black in the middle.

Field Notes: Associated with a variety of wetland and forest habitats.

Similar species: Most similar to *O. pulchella.*

Flowers: Meadow anemone and red elderberry in Minnesota.

Flight Dates: June - July

Status: Uncommon

Range: Northern

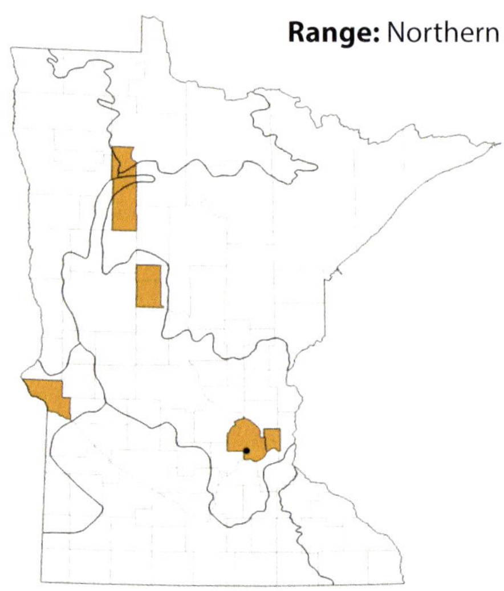

Genus *Myolepta* :: Spinelegs

The flies of the genus *Myolepta* are small, black-bodied flower flies. They occur in mature forests. Adults descend from the canopy to feed at flowers. The front and middle femurs have a number of small, black spines.

Scaled Spineleg

The Scaled Spineleg (***Myolepta strigilata***) is a small, Nearctic flower fly. The pile on the thorax is flattened and scale-like. The tibia are orange or yellow.

Field Notes: Larvae live in rot holes in deciduous trees.

Similar species: Most similar to *Myolepta nigra* which is also found in Minnesota. Also could be confused with other small, black flower flies.

Flowers: Common hoptree in Minnesota.

Flight Dates: June

Status: Rare

Range: Southern

Additional Minnesota Species Not Included in the Guide

Myolepta nigra: Eastern species. Black abdomen and black legs. No scale-like pile.

Genus *Neoascia* :: Fen Flies

The flies of the genus *Neoascia* are very small, possibly our smallest flower flies. The lower portion of the face protrudes forward, giving a duck-billed appearance in profile. The hind femurs are inflated. The larvae are associated with decaying leaf litter. The genus is divided into two subgenera: *Neoascia* and *Neoasciella*. **Neoascia** has sclerotized bridge behind metacoxa. *Neoasciella* has a gap behind the metacoxa.

Double-banded Fen Fly

The Double-banded Fen Fly (***Neoascia (Neoascia) metallica***) is a very small, Nearctic flower fly. It has two pair of yellow marks on the abdominal segments 3 and 4. Males have yellow distal tips to the hind femur.

Field Notes: Often observed at rewilding farm ponds.

Similar species: Very similar to *N. tenur*. *N. metallica* has yellow distal tips to the hind femur (black on *N. tenur*). Male genitalia must be examined to confirm identification. Could be confused with dark-colored *Sphegina* and also *Syritta*.

Flowers: Wild strawberry and dandelion in Minnesota.

Flight Dates: April - May

Status: Common

Range: Northern

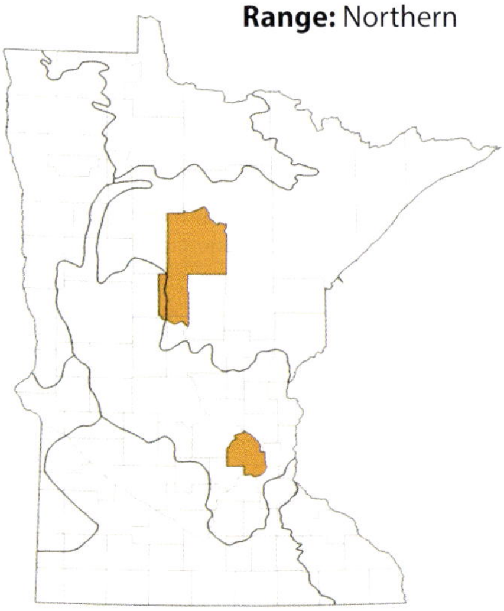

Neoascia metallica

Black-margined Fen Fly

The Black-margined Fen Fly (***Neoascia (Neoasciella) globosa***) is a very small, Nearctic flower fly. It has a single orange band on the third abdominal segment which doesn't extend all the way to the abdomen margin.

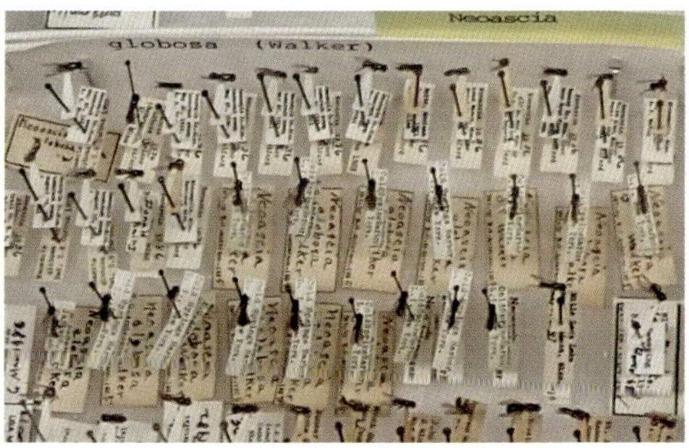

Range: Eastern

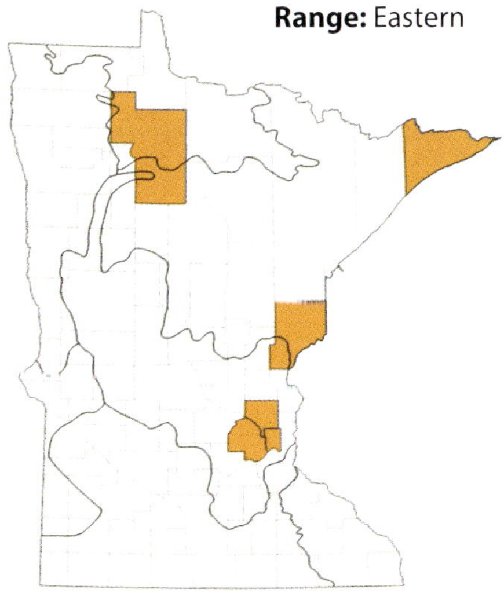

Field Notes: Associated with bogs.

Similar species: Most similar to other *Neoascia* and dark-bodied *Sphegina*.

Flowers: No floral records for Minnesota.

Flight Dates: April - May

Status: Uncommon

Additional Minnesota Species Not Included in the Guide

***Neoascia tenur; Neoascia subchalybea; Neoascia undescribed* sp.1** (in Skevington et. al.)

Genus *Sphegina* :: Pufftails

The flies of the genus *Sphegina* are small, wasp-waisted flower flies, often pale or brown in color. The hind femurs are inflated. The larvae are thought to live in decaying sap underneath bark. The genus is divided into two subgenera: *Sphegina* and *Asiosphegina. Sphegina* has, on the underside of the first abdomen segment, a dark, sclerotized plate, whereas *Asioshegina* does not. These flies, because of their small size and variable coloration, remain quite difficult to identify confidently.

Orange-horned Pufftail

The Orange-horned Pufftail (***Sphegina (Asiosphegina) campanulata***) is a small, Nearctic flower fly. It is distinguished by a pair of pollinose stripes on top of thorax. Flagellum is at least partially orange.

Similar species: Most similar to other species of *Sphegina* but could be confused with *Baccha*.

Field Notes: A hardwood forest species.

Flowers: Observed on goutweed, aniseroot, Canada honewort, and golden alexander in Minnesota.

Flight Dates: May - June

Status: Uncommon

Range: Eastern

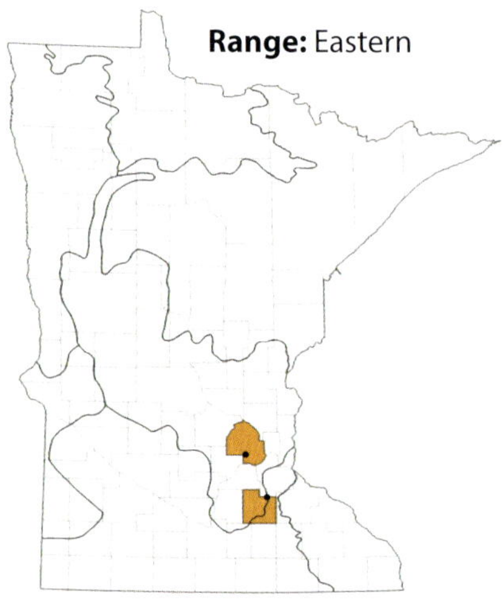

Black-horned Pufftail

The Black-horned Pufftail (***Sphegina (Asiosphegina) rufiventris***) is a small, Nearctic flower fly. A pair of faint pollinose stripes on top of a shiny, black thorax. Flagellum is dark-brown to black. The abdomen is often a deep, red color, as are the hind femurs.

Range: Eastern

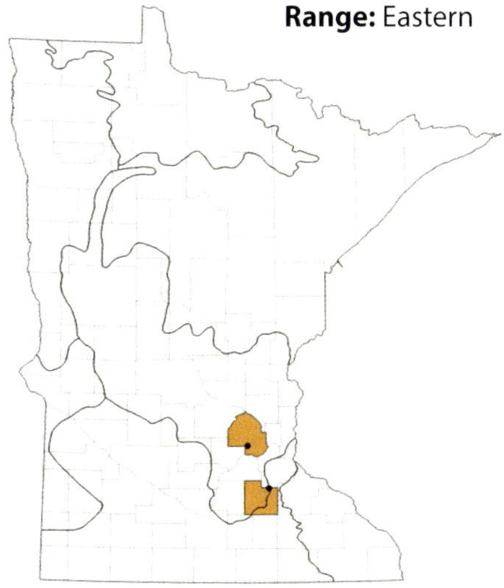

Field Notes: Known for mixed-boreal and bog habitats.

Similar species: Most similar to other species of *Sphegina* but could be confused with *Baccha*.

Flowers: Goldenrod and water hemlock in Minnesota.

Flight Dates: July - Sept

Status: Uncommon

Additional Minnesota Species Not Included in the Guide

Sphegina (Asiosphegina) petiolata: Northern; long spines on scutellum (*see images, right*).

Sphegina (Asiosphegina) brachygaster: Eastern; wide-waisted.

Sphegina (Asiosphegina) flavimana: Eastern; epandrium with turbercule.

Sphegina (Asiosphegina) lobata: Eastern; male 4th sternite asymmetrical and yellow-lobed.

Sphegina petiolata

Tribe Cerioidini

Tribe Cerioidini includes two closely related genera in Minnesota: *Ceriana* and *Sphixomorpha*.

Genus *Ceriana* :: Wasp Flies

The flies of the genus *Ceriana* are all convincing wasp mimics. They have antennae attached to a very long prominence. The flagellum is somewhat diamond-shaped and the arista is attached to the flagellum apically instead of dorsally. The larvae live in sap runs.

Northern Wasp Fly

The Northern Wasp Fly (***Ceriana abbreviata***) is a medium-sized, Nearctic flower fly. A convincing mason wasp mimic, for example, compare to wasps of the genus *Ancistrocerus*. The antennae are attached to a long prominence and the arista is attached to the end of the flagellum instead of dorsally. The coloration is usually black with yellow markings but the markings can sometimes be white. The face is yellow with black medial markings that form a "T".

Field Notes: The larvae of this fantastic fly live in sap runs on old or wounded trees. One fly was reared from larvae found on balsam poplar.

Similar species: Most similar to *Sphiximorpha willistoni* which is nearly identical except for the curiously different antennae build.

Flowers: Canada elderberry, cultivated parsnip, yarrow, and sedum in Minnesota.

Flight Dates: May - July

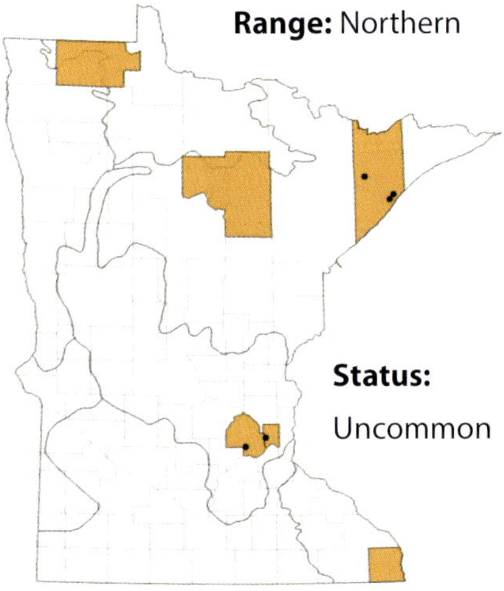

Range: Northern

Status: Uncommon

Genus *Sphixomorpha* :: Wasp Flies

The flies of the genus *Ceriana* are all convincing wasp mimics. They have long antennae attached to a short prominence. The flagellum is somewhat diamond-shaped and the arista is attached to the flagellum apically instead of dorsally. The larvae live in sap runs. (Sometimes treated as a subgenus of *Ceriana*)

Williston's Wasp Fly

Williston's Wasp Fly (***Sphiximorpha willistoni***) is a medium-sized, Nearctic flower fly. A convincing mason wasp mimic, for example, compare to wasps of the genus *Ancistrocerus*. The antennae have very long scapes which are attached to a short prominence and the arista is attached to the end of the flagellum instead of dorsally. The face is yellow with a black medial mark that resembles an arrow pointing downward. Larvae are associated with black walnut and poplar trees.

Range: Southern

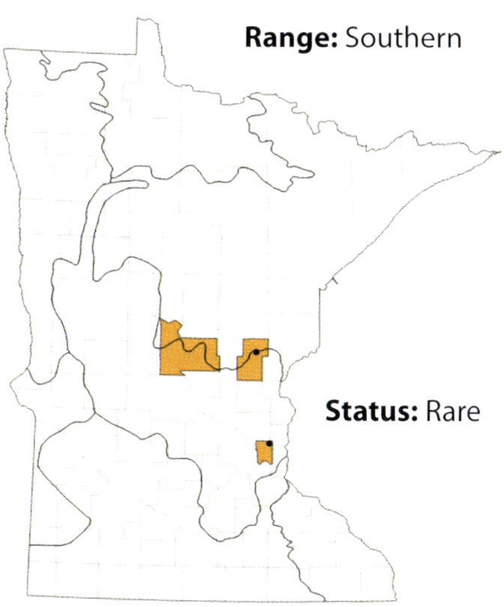

Status: Rare

Field Notes: This species is a known hilltopper, where the flies can be found perched on tree trunks.

Similar species: Almost identical to *Ceriana abbreviata* but differs in antennae build.

Flowers: No Minnesota floral records.

Flight Dates: May - July

Tribe Eristalini

Tribe Eristalini includes ten genera in Minnesota, further divided into three subtribes: Eristalina, Helophilina, and Sericomyiina.

Subtribe Eristalina > Genus *Eristalinus* :: Lagoon Flies

The flies of the genus *Eristalinus* are smallish, dark in color, and somewhat shiny. The wings are clear, without clouding. The R4+5 vein is deeply sinuous and wing cell r1 is closed, connected to wing margin by short vein. The eyes are light in color with an abundance of black, circular spots. They behave like muscid flies, perching on bare ground and stone. A single, non-native species has been recorded in Minnesota.

Common Lagoon Fly

The Common Lagoon Fly (***Eristalinus aeneus***), a medium-sized, Palearctic flower fly, is not common in Minnesota. This fly is dark-colored and shiny, sometimes with faint gray stripes on the dorsal surface of the thorax. The eyes are light colored with many dark, circular spots. The larvae feed on decaying plant matter.

Field Notes: In the United Kingdom, this fly is primarily a coastal species associated with ponds that have an abundance of rotting seaweed. A known hibernator in its native range, it's interesting that one of the Minnesota specimens was found overwintering in a cave.

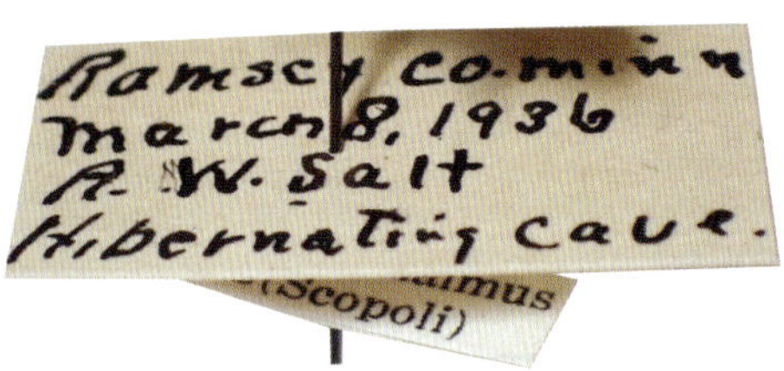

Range: Southern

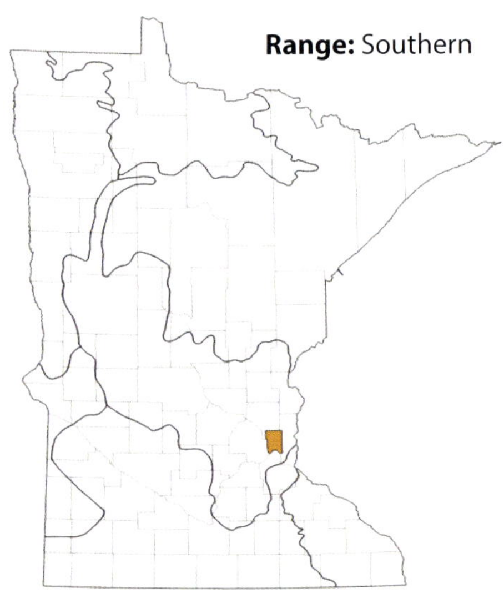

Similar species: Shouldn't be confused with any other flower fly.

Flowers: No records for Minnesota.

Flight Dates: March - April

Status: Rare. No modern records.

Subtribe Eristalina > Genus *Eristalis* :: Drone Flies

The genus *Eristalis* includes medium-sized and large flower flies. In most species, the face has a bare, vertical medial stripe. The arista on the antennae is usually bare but has visible hairs for a few species. The body is covered by fairly short hair, which makes the flies look somewhat honeybee-like. However, a few species have hair that is much longer and these resemble bumblebees. Wing coloration varies. The R4+5 vein is deeply sinuous and wing cell r1 is closed, connected to wing margin by short vein.

Drone flies live in a variety of habitats: wetlands, beaches, flood plains, sewage treatment ponds, feedlot runoff and other waters rich in organic matter. The larvae are able to take advantage of these nutrient rich but often anoxic habitats with the help of a long, extendable breathing tube. They are often referred to as rat-tailed maggots.

Adults visit a wide variety of flowers. Males exhibit territorial behavior, guarding flowers visited by females and guarding breeding sites.

There are 91 species worldwide. 20 species in North America; 11 species have been found in Minnesota. Known commonly as drone flies, some of these flower flies are the most abundant and most frequently encountered species.

Key to the Eristalis species

1 Eyes with a wide, double stripe of hair; very large, honey bee mimic (subgenus Eristalis) .... ***Eristalis tenax***

1′ Eyes with a single, less-pronounced stripe of hair or without stripes of hair (subgenus Eoseristalis) 2

2. Bumblebee mimic 3

2′ Unconvincing wasp or native bee mimic 4

3 Hind metatarsus reddish-orange; thorax and abdomen densely pilose .... ***Eristalis flavipes***

3′ Hind metatarsus black or dark brown; yellow face; thorax and abdomen densely pilose; orange, lateral marking on abdomen tergite usual distinct and visible .... ***Eristalis anthophorina***

4 Face completely covered in silverish-white hairs (though black underneath) 5

4′ Face with bare, black or orange vertical stripe 6

5 Middle leg basitarsi yellow; face entirely white pilose. Male eyes meet for a long distance .... ***Eristalis arbustroum***

5′ Middle leg tarsi entirely black or brown. Male eyes meet for a short distance .... ***Eristalis brousii***

6. Arista with visible setae along its entire length (microscope) 7

6′ Arista bare, at most a small grouping of 0-8 hairs dorsally near base (microscope) 9

7. Thorax striped transversely, making it appear bicolored (grey and black), scutellum bright yellow; abdomen black and bright yellow .... ***Eristalis transversa***

7′ Thorax uniformly dark or with indistinctly striped lengthwise, scutellum glossy 8

8 Black medial stripe on face. Spines on hind leg metatarsus all orange. Pterostigma well-defined and very narrow .... ***Eristalis interrupta***

8′ Orange medial stripe on face. Spines on hind leg metatarsus orange and black. Spot on wing; Antennae orange; abdomen lightly dusted .... ***Eristalis obscura***

9 Post alar callas black. Basal wing cell bc is darkened .... ***Eristalis dimidiata***

9′ Post alar callas yellow. Basal wing cell clear. Tergite hind margin light color; T4 with wavy tufts of hair .... ***Eristalis stipator***

Orange-spotted Drone Fly

The Orange-spotted Drone Fly (***Eristalis anthophorina***) is a large, Holarctic flower fly. A bumblebee mimic. Yellow pile on thorax and abdomen. Large orange spots on the sides of the second segment of the abdomen. These spots vary in size and intensity of color; they can also be hidden by dense pile on the abdomen. Wing crossveins are diffusely clouded. Hind tarsi are black. The larvae are of the rat-tailed maggot variety.

Field Notes: This somewhat squat drone fly tends to perch with its legs splayed.

Similar species: Most similar to *E. flavipes* but also could be confused with *Mallota*.

Flowers: Observed on dandelion, Labrador tea, aster, and goldenrod in Minnesota.

Flight Dates: May - September

Status: Uncommon

Range: Northern

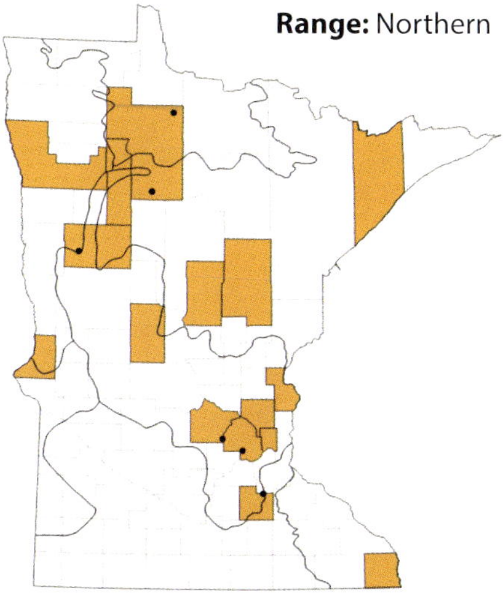

European Drone Fly

The European Drone Fly (***Eristalis arbustorum***) is a large, Palearctic species. It was introduced to North America in the late 1800s. Arista on antennae with long hairs (sparse compared to *Volucella*). The face is covered in white hair, though older, worn individuals and specimens may have a thin, black stripe. Eyes are pilose. For males, the yellow markings on the second and third sections of the abdomen define a medial shape: a black hourglass- or chalice-like shape. The yellow markings on females are reduced and limited to the first segment and define more of a thick black "X" though sometimes the females have no yellow markings at all. Wings are clear. The coloration of the middle leg must be examined carefully to distinguish this species from the closely-related *E. brousii*. The mesotibia is mostly yellow, black on the apical third. The basal tarsomere is also mostly yellow. This fly is a known migrant in its native range. Common to urban and agricultural areas. The larvae are of the rat-tailed maggot variety and develop in eutrophic water.

Field Notes: Introduced to North America in the late 1800s near Toronto and has subsequently spread across the entire continent, displacing the native species *E. brousii*.

Similar species: Most similar to *Eristalis brousii*.

Flowers: Hoary alyssum, goutweed, chrysanthemum.

Flight Dates: June - October

Range: Eastern

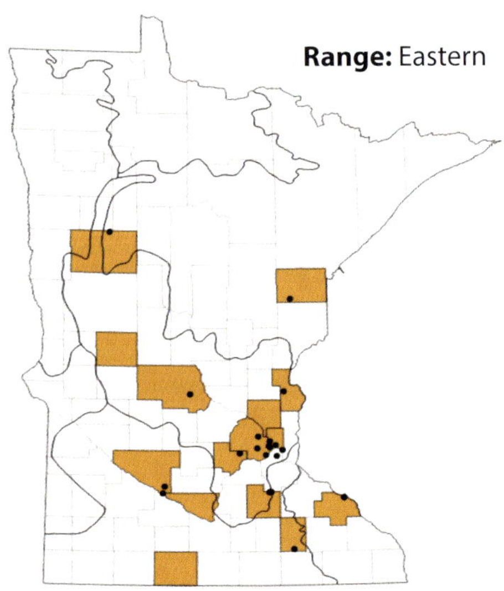

Status: Uncommon

Hourglass Drone Fly

The Hourglass Drone Fly (***Eristalis brousii***) is a large, Nearctic flower fly. This native fly is nearly identical to the non-native *E. arbustorum* described above. It differs in two observable characters: the coloration of the middle leg and the extent of contact of the eyes on male flies. The mesotibia is mostly black or dark, reddish-brown (*E. arbustorum* is mostly yellow and with a large amount of white hair). Male eyes meet for a short distance, less than the distance of the occiput (*E. arbustorum* eyes meet for a long distance, more than the distance of the occiput). The larvae are of the rat-tailed maggot variety.

Field Notes: The population of this species collapsed some decades after the introduction of the near-identical species, *Eristalis arbustorum*.

Similar species: Most similar to *Eristalis arbustorum*.

Status: Possibly extirpated in Minnesota though relict populations may still exist in the less-developed parts of the state. No modern records (1930s are the dates of the last collected specimens).

Flight Dates: June

Range: Northern

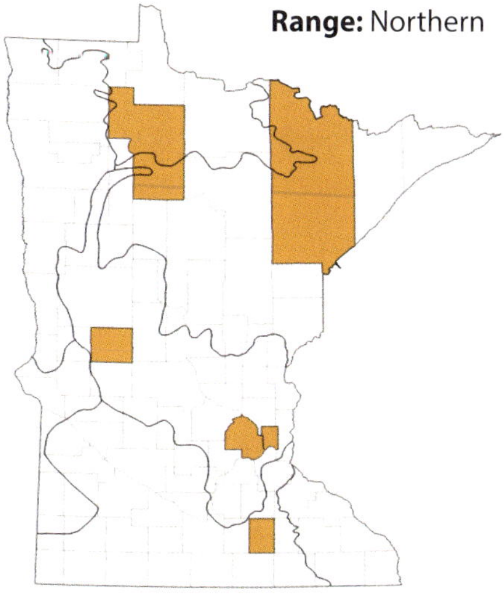

Flowers: No floral records for Minnesota.

Bog-dwelling Drone Fly

The Bog-dwelling Drone Fly (***Eristalis cryptarum***) is a large, Holarctic flower fly. This fly is recognized by the reddish-orange scutellum. The base and anterior edge of the wings are brownish-orange. It has a pair of dark-orange marks on the second segment of the abdomen. The abdomen segments are black with a white rim along the posterior edge of each segment. The flagellum is also, usually, orange in color. The larvae are of the rat-tailed maggot variety.

Field Notes: This species prefers oligotrophic wetlands such as bogs and fens.

Similar species: Superficially similar to other drone flies, but because of its unique color, it shouldn't be confused with these.

Flowers: No Minnesota floral records. Elsewhere, this fly is known to visit a variety of flowering plants associated with bog habitats, such as leatherleaf, bog rosemary, and Labrador tea.

Flight Dates: May

Range: Northern

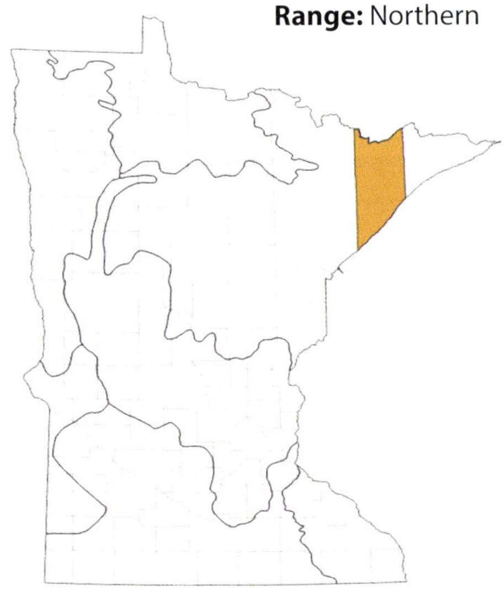

Status: Rare

Black-shouldered Drone Fly

The Black-shouldered Drone Fly (***Eristalis dimidiata***) is a large, Nearctic flower fly. An unconvincing bee mimic with shades of black-and-white wasp in the mix. Eyes with brown hairs. Male wings are clear except for the distinctive darkening of the basal wing cell bc. Female wings can be slightly darkened in other places. The antennae, facial stripe, thorax are black in color. The thorax has dense, short, whitish pile. The abdomen is black with thin, white abdominal rings at the end of each segment. Usually with a pair of yellow marks on segment two and a faded pair of marks on segment 3. Basal half of tibia are white, typically sharply divided white and black halves. The larvae are of the rat-tailed maggot variety and develop in eutrophic water.

Field Notes: Adults overwinter. Thus *E. dimidiata* are among the first flies encountered in early spring, often abundant on the first willow catkins.

Similar species: Most similar to *E. stipator* and the other dark-colored drone flies.

Flowers: Adults visit a wide variety of flowering plants, beginning with willow catkins and maple flowers in the spring, raspberry and sumac flowers in the summer, and the ubiquitous composites of autumn, the asters and goldenrods.

Status: Abundant

Range: Northern

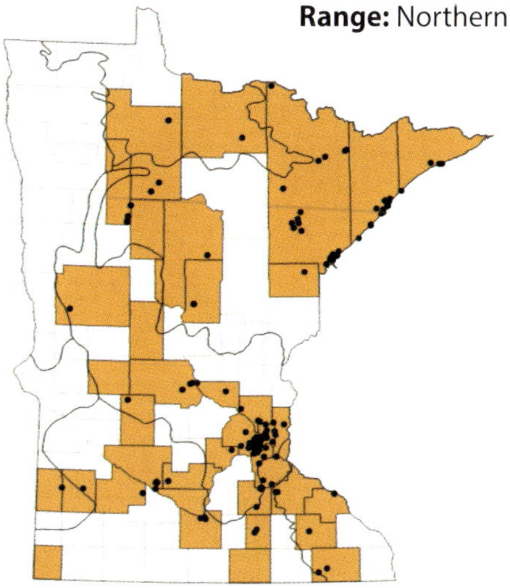

Flight Dates: March - November

Orange-legged Drone Fly

The Orange-legged Drone Fly (***Eristalis flavipes***) is a very large, Nearctic flower fly. A convincing bumblebee mimic. This species is easily recognized by its bright-orange metatarsus. The color of the abdomen pile can be highly variable, from mostly black, to yellow, to mostly orange. The larvae are of the rat-tailed maggot variety and develop in eutrophic water.

Field Notes: This convincing mimic not only looks like a bumble bee but behaves like one as well, flying slow through low vegetation just like a queen searching for a nesting site. Often observed ovipositing along shorelines of ponds and creeks.

Similar species: Most similar to *E. anthophorina* and other bumble bee mimics.

Flowers: Rattlesnake master, garlic chives, showy and Canada goldenrod, white snakeroot, meadowsweet, aster, and yarrow in Minnesota.

Range: Northern

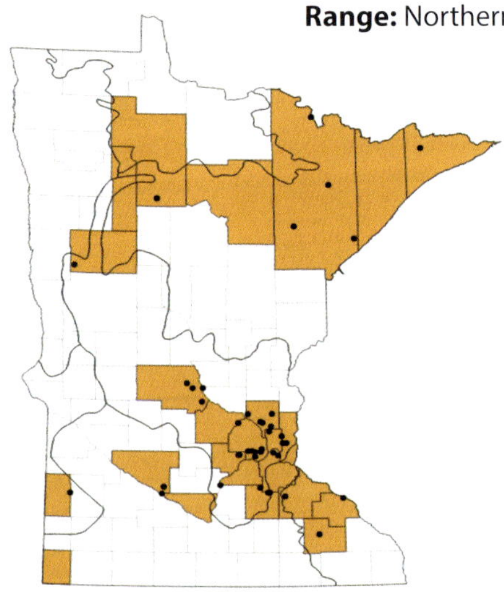

Status: Common

Flight Dates: May - October

Orange-spined Drone Fly

The Orange-spined Drone Fly (***Eristalis interrupta* = *Eristalis nemorum***) is a large, Holarctic species. The abdomen pattern is similar to several other drone flies with a pair of orange, triangular marks on the second abdominal segment. The medial facial stripe is black. The rear tibia is half dark, pale on the basal half. The rear basal tarsi are dark with orange spines. The wing has a small, sharply-defined pterostigma. The larvae are of the rat-tailed maggot variety.

Field Notes: Inhabits a variety of wetlands in the mixed boreal region of northern Minnesota.

Similar species: Most similar to *E. obscura* but could be mistaken for other drone flies.

Flowers: The sole floral record for Minnesota is yarrow.

Status: Rare

Flight Dates: July - August

Range: Northern

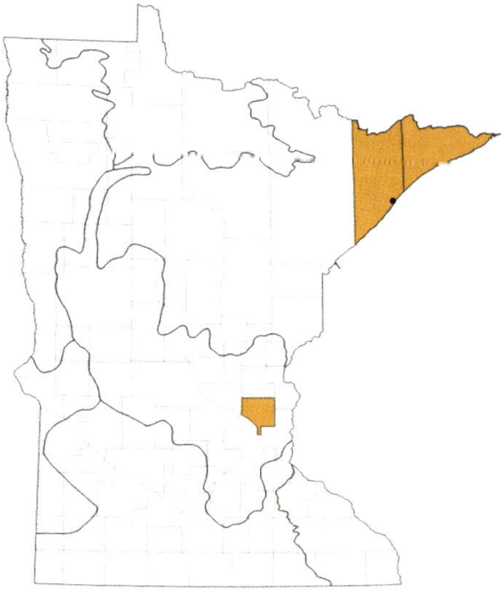

Dusky Drone Fly

The Dusky Drone Fly (***Eristalis obscura***) is a large, Nearctic flower fly. The abdomen pattern is similar to several other drone flies with a pair of orange, triangular marks on the second abdominal segment. The face has an orange medial stripe (black in other MN drone flies). The wing pterostigma is approximately square and slightly diffuse. The basal tarsi are pale with apical tarsi dark.

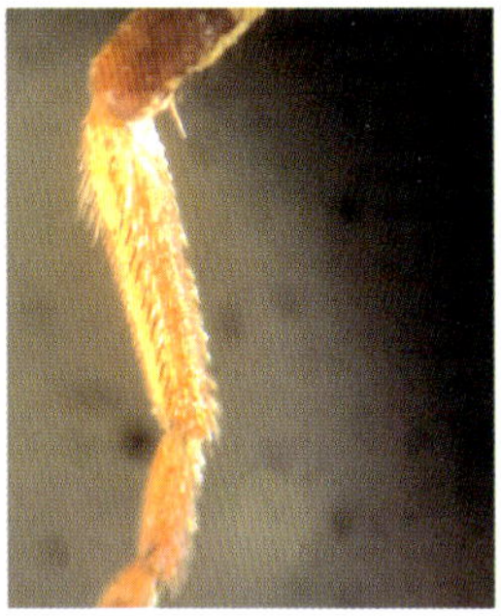

Field Notes: Inhabits a variety of wetlands in the mixed boreal region of northern Minnesota.

Similar species: Most similar to *Eristalis interrupta* but could be confused with other drone flies.

Flowers: Ox-eye daisy and yellow rocket in Minnesota. Elsewhere, these flies are known to visit a wide variety of flowering plants.

Flight Dates: May - July

Status: Uncommon

Range: Northern

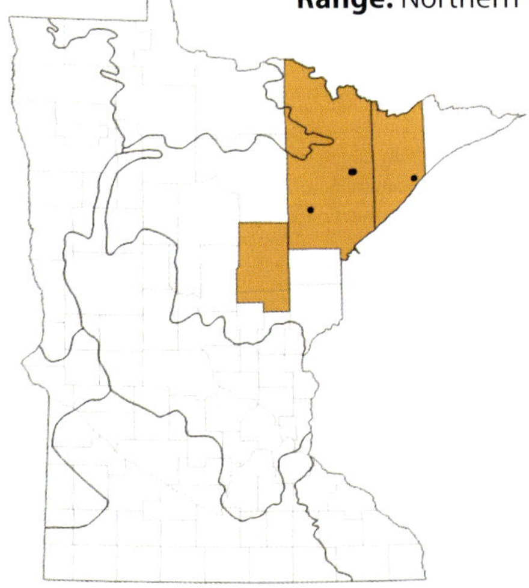

Yellow-shouldered Drone Fly

The Yellow-shouldered Drone Fly (***Eristalis stipator***) is a large, Nearctic flower fly. The abdomen pattern is similar to several other drone flies with a pair of orange, triangular marks on the second abdominal segment. The face has a black medial stripe. The pile on the postalar callus is yellow. The abdomen is shiny black with a white rim at the end of each segment. Most characteristic of this species, however, is the fringe of long, white pile on the fourth abdominal segment. The larvae are of the rat-tailed maggot variety and develop in eutrophic water.

Field Notes: The distribution of this fly is rather unusual in being common in the west and very rare in the east.

Similar species: Most similar to *Eristalis dimidiata* but could be confused with other drone flies.

Flowers: Adults visit a wide variety of flowering plants. In Minnesota records include goldenrod, creeping thistle, chrysanthemum, dandelion, obedient plant, aster, queen anne's lace, sedum, brown eyed susan, blanketflower, meadowsweet, clover, common milkweed, whorled milkweed, and wild plum.

Range: Western

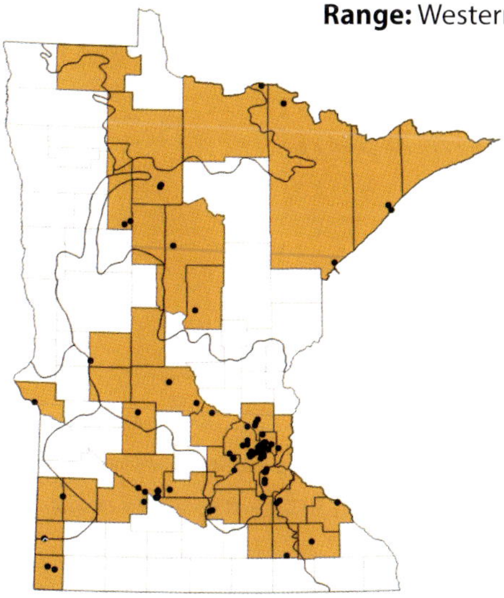

Status: Abundant

Flight Dates: May - October

Transverse-banded Drone Fly

The Transverse-banded Drone Fly (***Eristalis transversa***) is a large, Nearctic flower fly. The only black and yellow banded drone fly. The posterior margin of each abdominal segment is rimmed in yellow (white in other species) and a pair of yellow triangular marks on segment two. Another distinct character is the gray, transverse band on the black thorax. The larvae are of the rat-tailed maggot variety and develop in eutrophic water.

Field Notes: This fly is often heard before it's seen. Males basking on sunlit leaves, often at or above eye-level, are very territorial and will fly and drone loudly when disturbed or approached.

Flowers: In Minnesota floral records include sneezeweed, black-eyed susan, brown-eyed susan, cut-leaf coneflower, white snakeroot, asters, ox-eye daisy, sedum, sunflowers, blanket flower, dogwood, ninebark, and sumac. It's interesting that this yellow-marked drone fly seems to prefer yellow composites.

Flight Dates: May - October

Status: Abundant

Range: Eastern

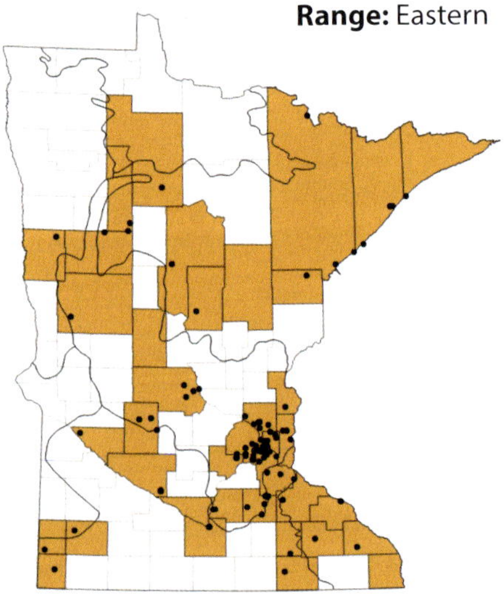

Similar species: Because of the unique color pattern of its abdomen, it shouldn't be confused with other species.

Common Drone Fly

The Common Drone Fly (***Eristalis tenax***) is a very large, Palearctic flower fly. This fly was introduced to North America in the 1800s. A convincing honey bee mimic. The hind legs are dark and curved. The tibia has long hairs that resemble the pollen basket of honeybees. The eyes have two vertical bands of hair. The face is long and has a wide, black medial stripe. The front tarsi are black. The larvae are of the rat-tailed maggot variety and develop in eutrophic water.

Field Notes: A known migrant in its native range. This fly overwinters in hibernaculums such as caves and houses. These overwintering flies survive only the warmest of Minnesota winters.

Similar species: The coloration and size of this fly should help distinguish it from other drone flies. It is, however, often mistaken for what it mimics, a honeybee.

Flowers: Adults visit a wide variety of flowering plants including, in Minnesota, dandelion, queen anne's lace, asters, mints, purple coneflower, and brown-eyed susan.

Range: Southern

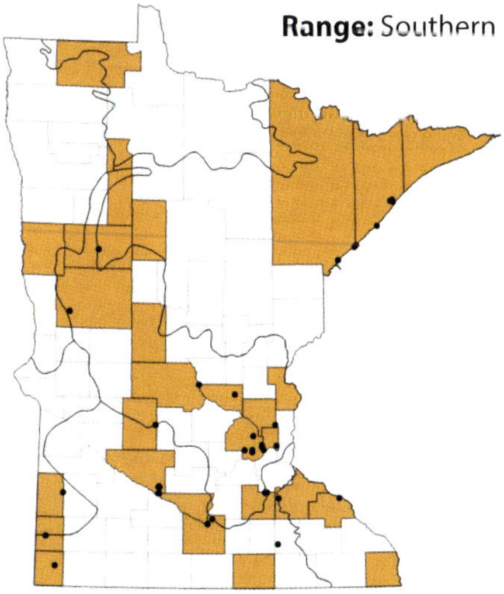

Status: Migrant

Flight Dates: August - November

Subtribe Helophilina > Genus *Arctosyrphus* :: Longbeaks

The genus *Arctosyrphus* is monotypic, including a single, Holarctic species. The R4+5 vein is deeply sinuous and the marginal cell r1 is open. The face is strongly produced downward as in the genera *Sericomyia* and *Volucella*. This single, Holarctic species has a unique distribution in North America with the majority of records coinciding with the prairie pothole region of the Great Plains. (*Arctosyrphus* was previously treated as a subgenus of *Lejops*.)

Northern Longbeak

The Northern Longbeak (***Arctosyrphus willingii)*** is a medium-sized, Holarctic flower fly. A very distinct species with a white-and-black face that is strongly produced downward and with a thick, whitish-gray pile on the thorax. The abdomen and legs are black. The larvae, of the rat-tailed maggot variety, were only recently described from specimens in Sakha Republic, in Russia.

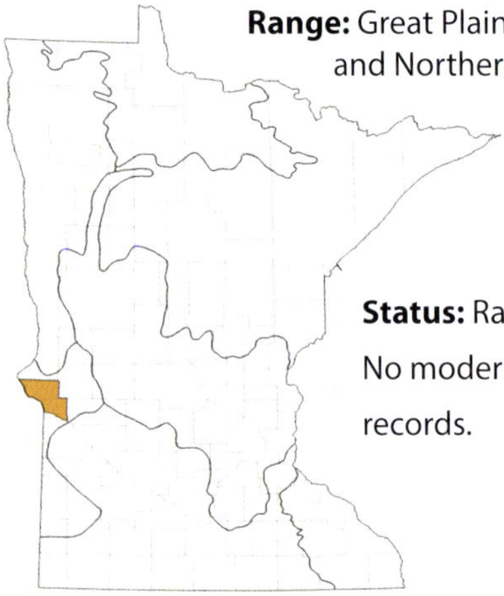

Range: Great Plains and Northern

Status: Rare. No modern records.

Field Notes: Boreal species. Curiously, a single record for Minnesota exists, not in the northern mixed-boreal part of the state but the southwest plains part of the state, in Big Stone County well over a century ago.

Similar species: Shouldn't be confused with any other flower fly.

Flowers: No records for Minnesota.

Flight Dates: No Minnesota dates. Elsewhere May - August.

Subtribe Helophilina > Genus *Anasimyia* :: Swamp Flies

The flies of the genus *Anasimyia* are a confusing group, no doubt in need of some taxonomic revision. The R4+5 vein is deeply sinuous and the marginal cell r1 is widely open. The pterostigma is narrow and well-defined. Males and females both have eyes that are separated. Despite that similarity, the species has a high degree of sexual dimorphism. Females have longitudinal thoracic stripes and similar abdomen patterns, pairs of pale swashes; Males vary in patterning and color, with abdomen marks being more hook-shaped. The larvae are aquatic, of the rat-tailed maggot variety. Three species are known from Minnesota. (*Anasimyia* was previously treated as a subgenus of *Lejops*.)

Two-lined Swamp Fly

The Two-lined Swamp Fly (***Anasimyia bilinearis***) is a medium-sized, Holarctic flower fly. The males are dark flies with reduced markings: segment two of the abdomen has wide, white markings on the sides; segment four has a faint pair of pollinose spots (and there is an even fainter pair on segment 3); scutellum is yellow. The female has a scutellum that is mostly yellow but outlined in black, a wide frons, and pairs of pale swashes on the abdomen.

Field Notes: This fly, often found in early spring around wooded wetlands, likes to perch on sunlit cattails and other winter-bleached debris.

Similar species: Males are distinctive. Females are difficult to separate from other *Anasimyia*.

Flowers: Wild plum, chokecherry, and buttercups in Minnesota.

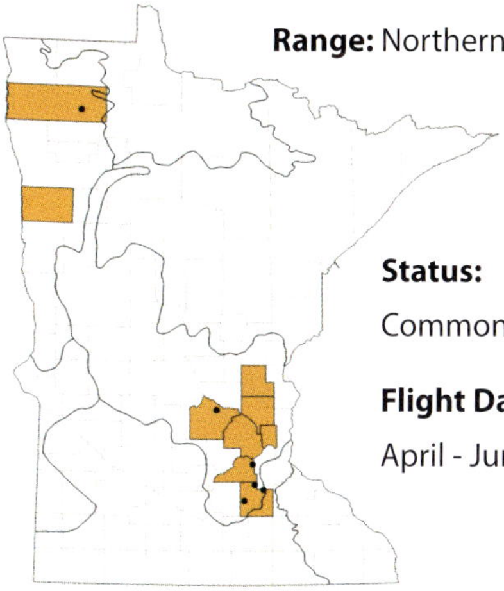

Range: Northern

Status: Common

Flight Dates: April - June

Lump-legged Swamp Fly

The Lump-legged Swamp Fly (***Anasimyia chrysostoma***) is a medium-sized, Nearctic flower fly. The scutellum is more than half black. The males have narrow abdomens that are often brightly colored, with white markings and a reddish-orange wash. The rear tibia is curved with an apical spine. The female frons is narrow.

Field Notes: Males are aggressively territorial and can sometimes be observed fighting.

Similar species: The narrow abdomen and reddish-orange color of males are distinct.

Flowers: Fleabane is the only floral record recorded for Minnesota so far.

Flight Dates: June

Status: Uncommon

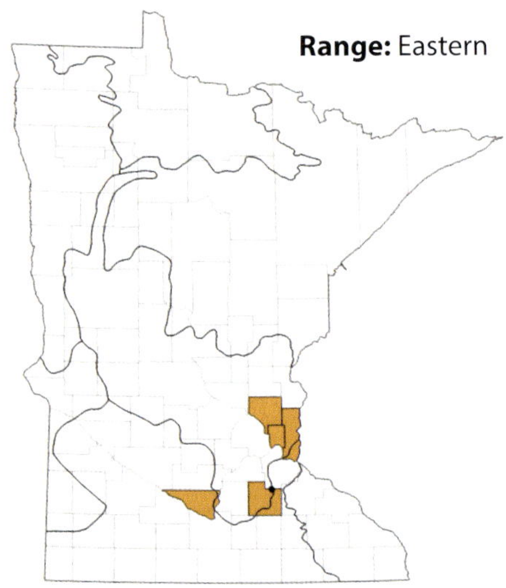

Range: Eastern

Additional Minnesota Species Not Included in the Guide

Anasimyia anausis: Northern; Males and females similarly patterned and colored: yellow scutellum; pale swashes on black abdomen.

Subtribe Helophilina > Genus *Eurimyia* :: Swamp Flies

The flies of the genus *Eurimyia* were formerly treated as a subgenus of *Lejops*. Three species are recognized worldwide. Of these, one is found in North America and Minnesota. The R4+5 vein is deeply sinuous and the marginal cell r1 is widely open. The pterostigma is small and diffuse. Males and females both have eyes that are separated. Most characteristic, however, is the pointy face which protrudes forward in the shape of a cone. Three species are known from Minnesota.

Long-nosed Swamp Fly

The Long-nosed Swamp Fly (***Eurimyia stipata***) is a medium-sized Nearctic species, a sister species of the Palearctic species Eurimyia lineata. This fly is recognized by its cone-shaped face which extends farther than the antennae. Sexually dimorphic coloration with the male having wider, orange markings on the abdomen. **Note:** The taxonomy of this species is a little messy. In Europe, it's *Anasimyia lineatus*. Here it's *Eurimyia stipata* though it used to be *Lejops (Eurimyia) lineatus*. The common name "Long-nosed Swamp Fly" sidesteps the thicket of Latin.

Field Notes: Usually found near water, often perches on emergent vegetation.

Similar species: The Two-lined Swamp Fly and the Moon-shaped Swamp Fly are very similar in coloration and markings. It could also be confused with some of the Bog Flies (genus *Parhelophilus*).

Flowers: Sweetflag iris, spiderwort, creeping thistle, and dandelion in Minnesota.

Flight Dates: May - June

Range: Northern

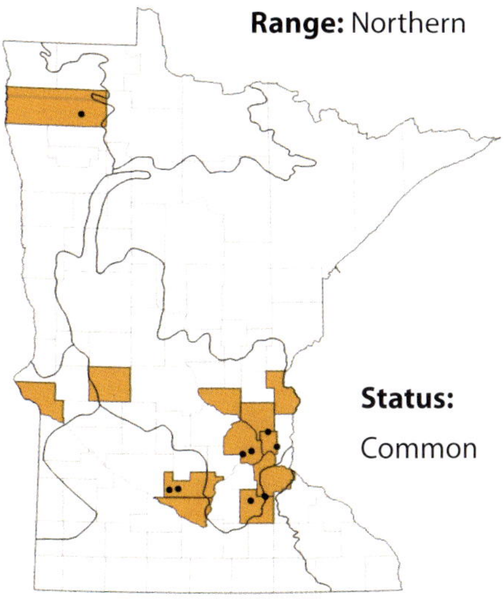

Status: Common

Subtribe Helophilina > Genus *Helophilus* :: Marsh Flies

The flies of the genus *Helophilus* are large, showy, yellow-and-black, not-overly-convincing wasp mimics. They included some of our most likely to be encountered flies, pretty much anywhere there is water and bright sunshine. The R4+5 vein is deeply sinuous and the marginal cell r1 is narrowly open. The pterostigma is long and diffuse. Males and females both have eyes that are separated. They are common at almost all aquatic habitats and can be particularly abundant at small ponds and puddles with decaying organic matter.

Etymology: *helo+philus* = marsh-loving

Key to the Helophilus species

1	Orange median stripe on face	2
1′	Black median stripe on face	3
2	The scape and pedicel of the antennae are orange. Males: the space between the eyes is narrow, less than the width of the antennal pits; Females: the space between the ocelli and the antennae is black pilose	***Helophilus fasciatus***
2′	Males: the space between the eyes is wider than the width of the antennal pits; Females—the space between the ocelli and the antennae is yellow pilose	***Helophilus latifrons***
3	Thoracic stripes reduced	4
3′	Thoracic stripes broad	***Helophilus hybridus***
4	Wide yellow marks on abdominal section 3	***Helophilus obscurus***
4′	Narrow yellow marks on abdominal section 3	***Helophilus lapponicus***

Narrow-headed Marsh Fly

The Narrow-headed Marsh Fly (***Helophilus fasciatus***) is a large, Nearctic flower fly. Four parallel, pollinose stripes on the thorax. The medial facial stripe is yellow. Abdominal segments 2 and 3 have large yellow markings. The scape and pedicel of that antennae are orange. Male eye separation is narrow, about equal to the width between the two middle thoracic stripes.

Field Notes: This common, colorful fly can be encountered at small ponds and puddles with decaying organic material. The females lay eggs on vegetation overhanging water. When the eggs hatch the larvae drop into the water.

Similar species: Most similar to *H. latifrons*. All *Helophilus* species look alike and require careful examination.

Flowers: Willow catkins, aster, creeping thistle, tansy, goldenrod, sedum, chokecherry, wild rose, wild plum, forget-me-not, sunflower, purple coneflower, yarrow, white snakeroot, dandelion, Virginia strawberry, and pasqueflower.

Range: Eastern

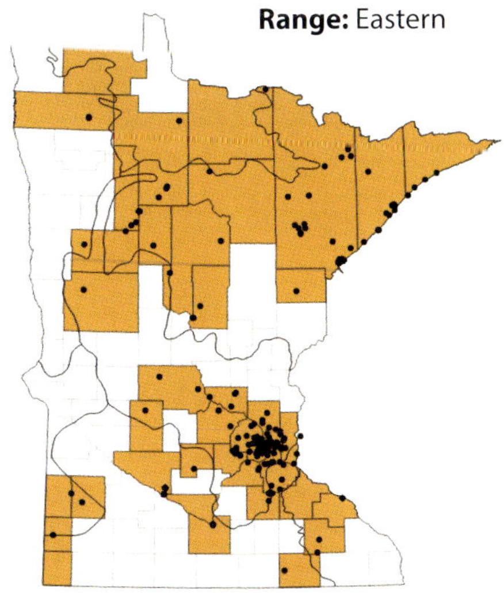

Status: Abundant

Flight Dates: April - October

Broad-headed Marsh Fly

The Broad-headed Marsh Fly (***Helophilus latifrons***) is a large, Nearctic flower fly. Four parallel, pollinose stripes on the thorax. The medial facial stripe is yellow. Male abdominal segments 2 and 3 have large yellow markings, more geometric than rounded, and bright yellow. Females have thin, curved markings on segments 3 and 4. The scape and pedicel of that antennae are black. Male eye separation is wider than *H. fasciatus*, slightly greater than the width between the two middle thoracic stripes.

Field Notes: Populations of this fly may be declining.

Similar species: Most similar to *H. fasciatus*. All *Helophilus* species look alike and require careful examination.

Flowers: Goldenrod, hoary alyssum, boneset, and aster in Minnesota.

Flight Dates: June - September

Status: Common

Range: Eastern

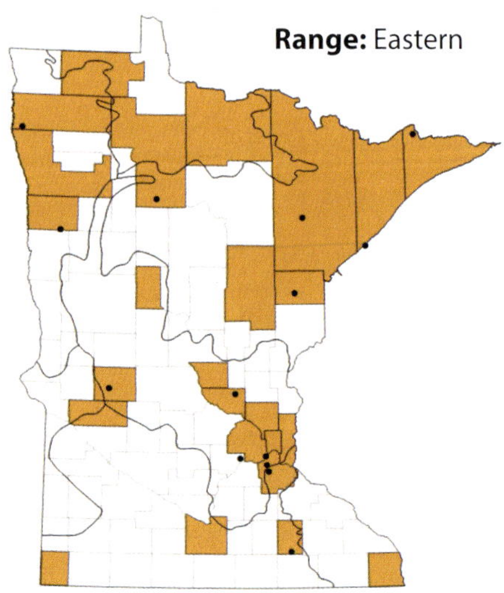

Woolly-tailed Marsh Fly

The Woolly-tailed Marsh Fly (***Helophilus hybridus***) is a large, Holarctic flower fly. Four parallel, pollinose stripes on the thorax. The medial facial stripe is black. Females' abdominal segments 2 and 3 have large yellow markings. The scape and pedicel of that antennae are black. Females' abdominal segment 5 is mostly yellow on the margin. Males have a tuft of long pile at the end of the abdomen.

Field Notes: One specimen (above, right) had the large pollen grains of Milkweed stuck to its feet.

Similar species: Most similar to *H. obscurus*. All *Helophilus* species look alike and require careful examination.

Flowers: Common Milkweed in Minnesota.

Flight Dates: July - September

Status: Uncommon

Range: Western

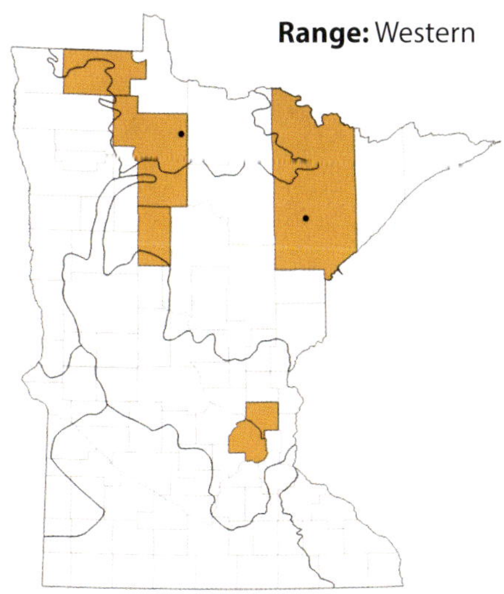

Additional Minnesota Species Not Included in the Guide

Helophilus obscurus: Northern; Black medial stripe on face. Reduced thoracic stripes. Wide yellow marks on T3.

Helophilus lapponicus: * Northern; Black medial stripe on face.

Subtribe Helophilina > Genus *Parhelophilus* :: Bog Flies

The flies of the genus *Parhelophilus* most resemble those of the genus *Helophilus* but are much smaller, have a distinct orange color, and differ subtly in wing venation. The R4+5 vein is deeply sinuous and the marginal cell r1 is widely open. The pterostigma is narrow and dark. Males and females both have eyes that are separated. The larvae have long breathing tubes, of the rat-tailed maggot variety, allowing them to live in the low-oxygen waters associated with decaying organic matter.

Etymology: *par+helophilus* = near *Helophilus* or similar to *Helophilus*

Common Bog Fly

The Common Bog Fly (***Parhelophilus laetus***) is a medium-sized, Nearctic flower fly. The R4+5 vein is deeply sinuous and the marginal cell r1 is widely open. The pterostigma is narrow and dark. These flies have an overall orange appearance. The face is entirely orange. Pollinose thoracic stripes (often there is an additional, thin central stripe). Males and females both have eyes that are separated. Male specimens can be confirmed under the microscope by their long cerci. Rear tibia is orange in the middle; the femur is orange basally.

Field Notes: This fly doesn't venture very far from the water's edge, where it's often encountered perched on sedge and grass blades.

Similar species: Most similar to P. *divisus* and *P. integer* both eastern species not known from Minnesota.

Flowers: *Carex*

Status: Common

Flight Dates: May - August

Range: Eastern, with scattered records in the west.

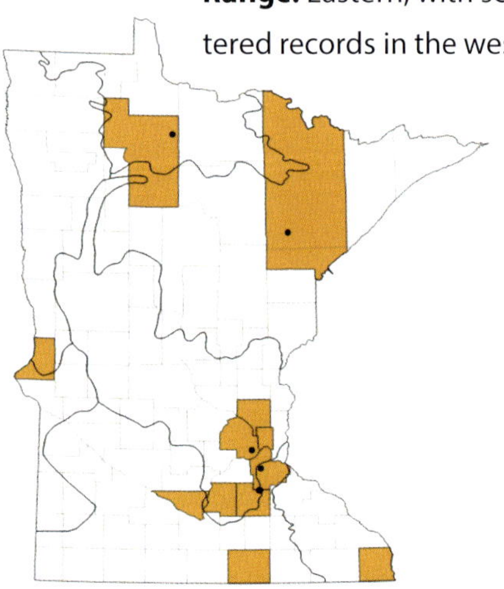

Dusky Bog Fly

The Dusky Bog Fly (***Parhelophilus rex***) is a medium-sized, Nearctic flower fly. The R4+5 vein is deeply sinuous and the marginal cell r1 is widely open. The pterostigma is narrow and dark. These flies have a black and yellow appearance. Males have a pair of very thin thoracic stripes that thicken posteriorly. Often, there's an additional, thin, central stripe. The hind femur is black at the base.

Field Notes: The next time you encounter yellow water lilies while canoeing or boating stop and look inside the flowers. Most likely you'll encounter some of these flies there.

Similar species: Most similar to other *Parhelophilus* species, but also similar in appearance to *Helophilus*.

Flowers: Willow catkins , yellow water lilies, and dandelion.

Flight Dates: June - August

Range: Northern

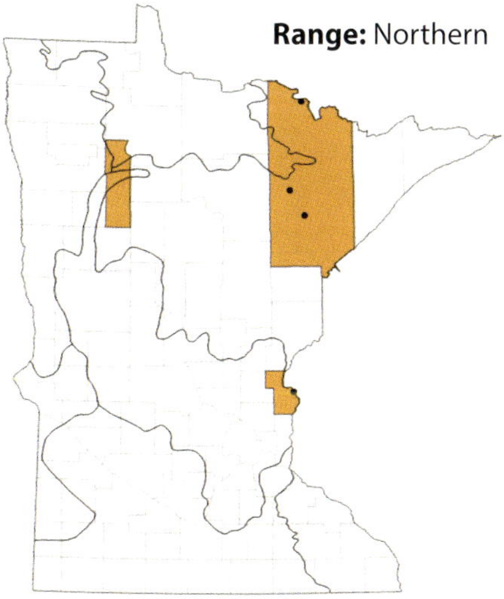

Status: Common

Additional Minnesota Species Not Included in the Guide

Parhelophilus brooksi: Northern; thin black abdomen with yellow legs

Parhelophilus obsoletus: Northern; no thoracic stripes and orange face and abdomen markings

Parhelophilus porcus: Northern; wide black abdomen with gray spots

Subtribe Helophilina > Genus *Polydontomyia* :: Sicklelegs

The genus *Polydontomyia* is monotypic. The one species is sexually dimorphic, so much so that the males and females look remarkably different. The males are easily recognized by their red abdomens; the females are a greenish-gray color with faint white markings on the abdomen. Both males and females have strongly curved hind legs, the tibia with a sharp spur on the apical end. Like other genera in the subtribe helophilina the R4+5 vein is deeply sinuous and the marginal cell r1 is open. The pterostigma is narrow and dark. Males and females both have separated eyes.

Dimorphic Sickleleg

The Dimorphic Sickleleg (***Polydontomyia curvipes***) is a large, Holarctic flower fly. The R4+5 vein is deeply sinuous and the marginal cell r1 is open. The pterostigma is narrow and dark. The males are easily recognized by their red abdomens; the females are a greenish-gray color with faint white markings on the abdomen. Both males and females have strongly curved hind legs, the tibia with a sharp spur on the apical end.

Field Notes: A specialist of brackish wetlands. It should come as no surprise that the one persistent Minnesota population has been at Salt Lake Wildlife Management Area at the western border of the state.

Similar species: Shouldn't be confused with other flower flies.

Flowers: No floral records for Minnesota.

Status: Rare

Flight Dates: August

Range: East and West Coast plus the brackish potholes of the upper Great Plains.

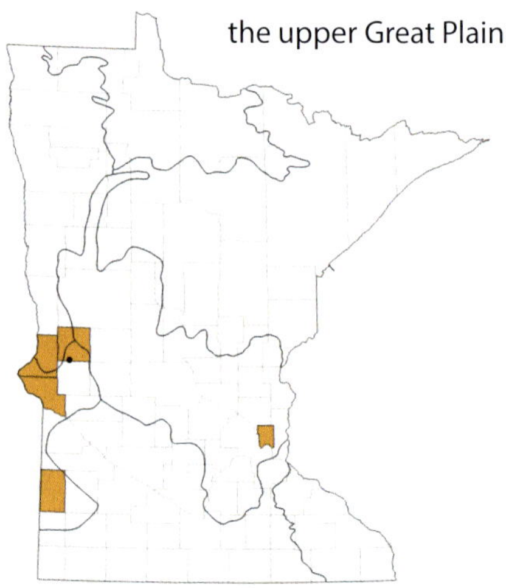

Subtribe Helophilina > Genus *Mallota* :: Mimics

The flies of the genus *Mallota* are convincing bumblebee mimics. The R4+5 vein is deeply sinuous and the cell r1 is open. Enlarged hind tibia and femur. Male eyes meet dorsally. The larvae, of the rat-tailed maggot variety, live in rot holes and deep, water-filled hollows in older trees. Two species are known from Minnesota with a third species a likely possibility.

Hairy-eyed Mimic

The Hairy-eyed Mimic (***Mallota posticata***) is a large Nearctic flower fly. A convincing bumblebee mimic. The eyes are pilose (note that the hairs are most visible on the lower parts of the eyes so the eyes might appear bare in dorsal photos). The Thorax, scutellum, and abdomen are covered in thick yellow pile. The extent of the pilosity on the abdomen is variable.

Field Notes: The larvae, of the rat-tailed maggot variety, live in rot holes and deep, water-filled hollows in older trees.

Similar species: Most similar to *Mallota bautias* and other bumblebee mimics, including the *Laphria* robber flies.

Flowers: Mock orange in Minnesota. Most often observed sitting on leaves.

Flight Dates: June - July

Range: Eastern

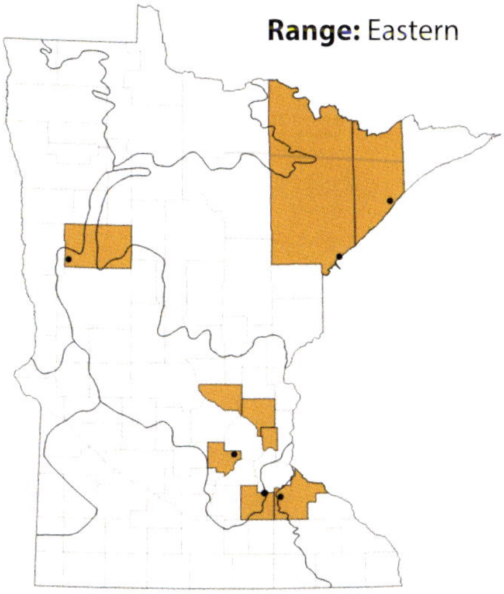

Status: Common

Bare-eyed Mimic

The Bare-eyed Mimic (***Mallota bautias***) is a large Nearctic flower fly. A convincing bumblebee mimic. The eyes are bare. The thorax and scutellum are covered in thick, yellow pile. The abdomen is black pilose.

Field Notes: The larvae, of the rat-tailed maggot variety, live in rot holes and deep, water-filled hollows in older trees.

Similar species: Most similar to *Mallota posticata* and other bumblebee mimics, including the *Laphria* robber flies.

Flowers: Fleabane, mustard, wild rose, and ninebark in Minnesota.

Flight Dates: May - July

Range: Eastern

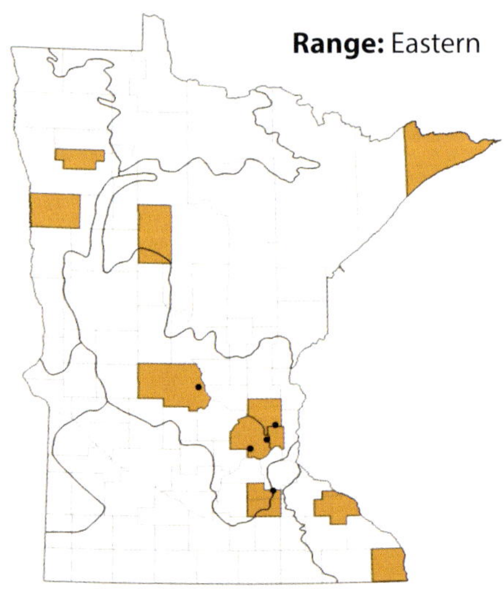

Status: Common

Additional Minnesota Species Not Included in the Guide

Mallota illinoensis: * Southern; bare eyes; large brown spot on wings; longer body.

Subtribe Sericomyiina > Genus *Sericomyia* :: Pond Flies

The flies of the genus Sericomyia are large, often colorful, imperfect mimics of social wasps. They can be recognized by their plumose arista. The R4+5 vein is usually sinuous (but in a few of the species it is nearly straight) and the marginal cell r1 is open. The pterostigma is long and diffuse like that of *Helophilus* but it is often obscured by the coloration of the wings along the foredge. Male eyes meet dorsally. The larvae are aquatic, of the rat-tailed maggot variety.

Yellow-spotted Pond Fly

The Yellow-spotted Pond Fly (***Sericomyia transversa***) is a large, Nearctic flower fly. An unconvincing wasp mimic. The second abdominal segment has a pair of large white spots. the third and fourth segments have a pair of thick, white bars. The legs are mostly black.

Field Notes: The yellow-spots are usually white-spots in Minnesota. A bog species.

Similar species: Most similar to *Sericomyia lata.*

Flowers: Labrador tea in Minnesota. Elsewhere observed visiting meadowsweet and wild plum.

Flight Dates: June

Status: Rare

Range: Eastern

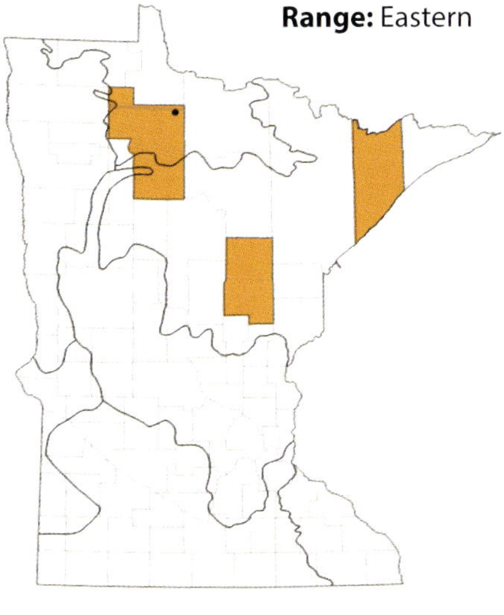

White-spotted Pond Fly

The White-spotted Pond Fly (***Sericomyia lata***) is a large, Nearctic flower fly. An unconvincing wasp mimic. The second abdominal segment has a pair of large white spots. The third and fourth segments have a pair of thin, white bars that are constricted in the middle and often divided in two. The face is yellow and the legs are mostly black.

Field Notes: The end of the abdomen is often a reddish-orange color on males.

Similar species: Most similar to *Sericomyia transversa*.

Flowers: Chokecherry, anemone, mustard, and dandelion in Minnesota.

Flight Dates: June

Status: Uncommon

Range: Northern

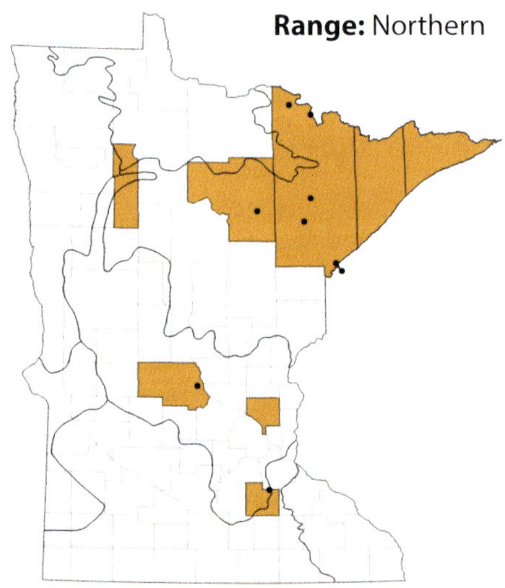

Oblique-banded Pond Fly

The Oblique-banded Pond Fly (***Sericomyia chrysotoxoides***) is a large, Nearctic flower fly. The R4+5 vein is nearly straight. The abdominal sections 2, 3, and 4 have pairs of thin yellow bands that angle upwards medially. The legs are mostly pale.

Range: Eastern

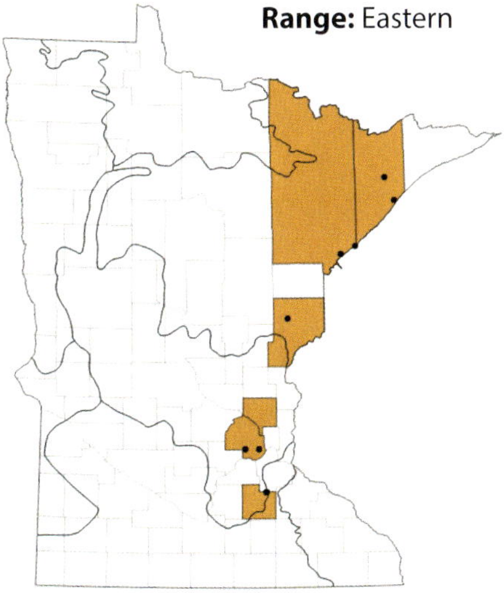

Field Notes: The most common pond fly in Minnesota. Often found basking in the sun with wings open wide.

Similar species: Most similar to *Sericomyia bifasciata* and is often mistaken for *Chrysotoxum* species, though the antennae differ dramatically.

Flowers: Aster, goldenrod, and black cohosh in Minnesota.

Flight Dates: July - October

Status: Common

Narrow-banded Pond Fly

The Narrow-banded Pond Fly (***Sericomyia militaris***) is a large, Nearctic flower fly. A convincing social wasp mimic. The thorax, scutellum, and abdomen are dark. The abdomen markings are variable but are most often pairs of very thin white bands that are slightly oblique, and often absent on segment two. The hind tibia and tarsi are orange.

Field Notes: Most abundant in the mixed-boreal region of the state.

Similar species: Most similar to our other bald-faced hornet mimic *Spilomyia fusca.*

Flowers: Bunchberry, ox-eye daisy, dandelion, pearly everlasting, aster, and goldenrod in Minnesota.

Flight Dates: June - October

Status: Common

Range: Northern

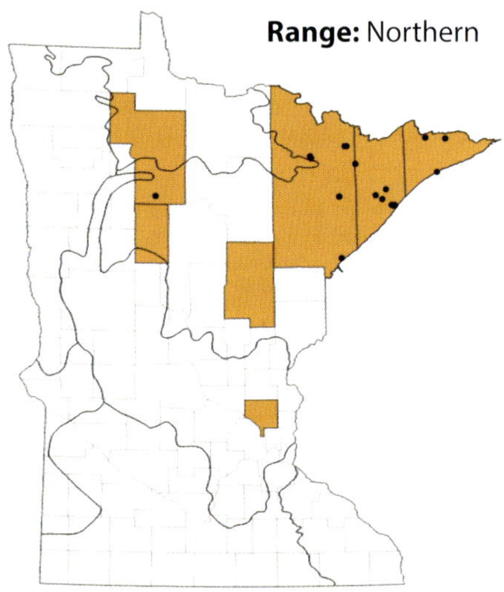

Additional Minnesota Species Not Included in the Guide

Sericomyia bifasciata: * Eastern; males with two pair of yellow bands on abdomen

Sericomyia sexfasciata: * Northern; six widely-separated abdomen markings that are slightly concave

Sericomyia vockerothi: Northern; a bumblebee mimic with yellow thorax and black abdomen.

Tribe Merodontini

Tribe Merodontini includes two genera in Minnesota: *Eumerus* and *Merodon*. The larvae of these flies feed in flower bulbs.

Genus *Merodon* :: Bulb Flies

The flies of genus *Merodon* are large bee mimics. The R4+5 vein is deeply sinuous and the marginal vein M1 bends back toward the base. There is a single, introduced species in North America.

Narcissus Bulb Fly

The Narcissus Bulb Fly (***Merodon equestris***) is a large, non-native flower fly. This European species was introduced to North America at some point after 1900 through the importing of flower bulbs. A polymorphic, bumble bee mimic. The species is identified by wing venation and black legs and the thick, hind femur that has a triangular spur apically. Populations are probably centered around large urban areas where there are more flower gardens and more bulbs upon which the larvae feed.

Field Notes: Two rather distinct color morphs exist as if the species can't decide which bumblebee species to mimic. Up to seven varieties have been recognized and given names in Europe. Two of these varieties seem prevalent in Minnesota: the narcissi form (fuzzy orange thorax and pale-haired abdomen) and the equestris form (yellow and black thorax).

Similar species: Most likely to be confused with *Pterallastes* , but also could be confused with flies of the genus *Mallota* as well as the other bumblebee mimics.

Flowers: Adults have been observed on a variety of flowers, including Mock Orange, Wild Geranium, Yarrow, Dandelion, and Viburnum.

Flight Dates: May - June

Status: Uncommon

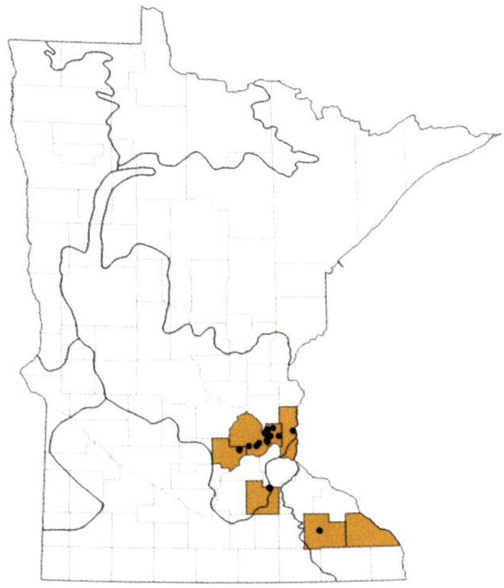

Range: Introduced species with large east and west coast populations.

Genus *Eumerus* :: Bulb Flies

The flies of genus *Eumerus* are small dark flies with faint pollinose marks on the abdomen. The marginal wing vein, M1, is biangulate ("W"-shaped) with small spurs at the apex of each of the two bends. Three species have been introduced to North America. And, so far, only a single species has been recorded in Minnesota.

Lesser Bulb Fly

The Lesser Bulb Fly (***Eumerus funeralis***) is a small, Palearctic flower fly. A small, black fly. Faint, white pollinose stripes on the thorax. The abdomen markings consist of pairs of faint, oblique dashes. The face is yellow pilose. The male has a small tubercle on the underside of the rear femur basally. The larvae are saprophagous and feed on flower bulbs that are damaged.

Field Notes: More common in urban and developed areas, especially locations with flower gardens with bulbs.

Similar species: Most similar to *Eumerus strigatus*. It could also be mistaken for *Orthonevra* or *Chrysogaster*.

Flowers: The single floral record in Minnesota so far is strawberry. Other observations have been on leaves near the fly's host plants.

Flight Dates: June - August

Range: Southern

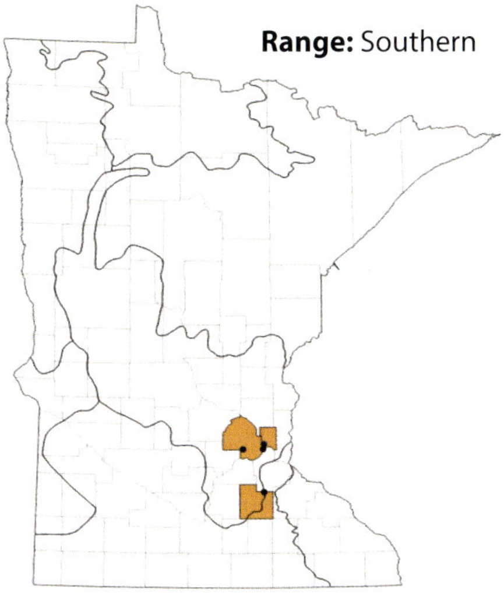

Status: Uncommon

Additional Minnesota Species Not Included in the Guide

Eumerus strigatus: * Northern; white pilose face; no tubercle on the rear femur.

Tribe Milesiini

Tribe Milesiini includes fourteen genera in Minnesota, further divided into six subtribes: Blerina, Criorhinina, Milesiina, Temnostomina, Tropidiina, and Xylotina.

Subtribe Blerina > Genus *Somula* :: Wood Flies

The genus *Somula* contains only two species, both found in North America. They are large, yellow-and-black flies that are convincing social wasp mimics. Only one of the two species is known from Minnesota.

Spotted Wood Fly

The Spotted Wood Fly (***Somula decora***) is a large, Nearctic flower fly. The markings on the abdomen are unique: pairs of large yellow marks that are boldly outlined in black. The marks on segment two of the abdomen are curved and look something like a pair of lazy commas. The face is yellow and has a noticeable prominence to which the antennae are attached. The legs are thin and mostly yellow. The wings are darkened along the foredge. The larvae develop in decaying trees.

Field Notes: This eye-catching fly has been observed ovipositing in the hollow at the base of a red oak in Minnesota.

Similar species: Most similar to other large, yellow-and-black social wasp mimics, such as *Temnostoma* and *Spilomyia.*

Flowers: Sumac is the only floral record for Minnesota so far. Elsewhere this fly has been observed on dogwood and hawthorn.

Flight Dates: May - July

Range: Eastern

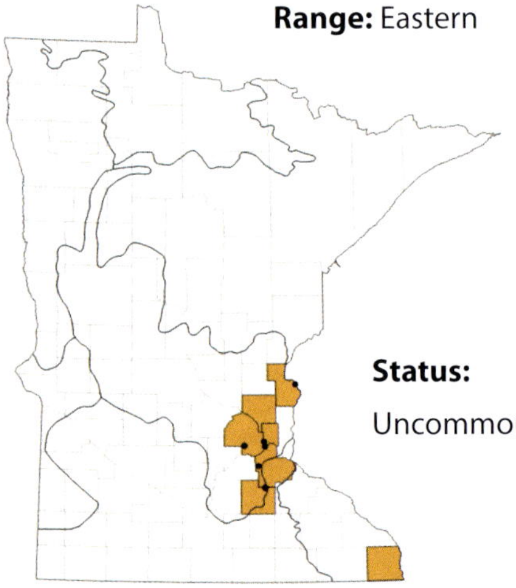

Status: Uncommon

Subtribe Blerina > Genus *Blera* :: Wood Flies

The flies of the genus *Blera* are black, wide-bodied flies. The head is equal in width to the thorax. And the abdomen is usually as wide or wider than the thorax. The R4+5 vein is straight. The crossvein r-m connects to cell dm at its middle. The face has a significant prominence to which the antennae attach. The larvae feed on decaying organic matter, usually in stumps and roots of old trees, both conifers and deciduous. Three species are known from Minnesota.

Orange-tailed Wood Fly

The Orange-tailed Wood Fly (***Blera analis***) is a large, Nearctic flower fly. A black, wide-bodied fly. The face and antennae are orange. Segments 2 and 3 of the abdomen are black. Segments 4 and 5 are orange. The hind femur is black.

Range: Eastern

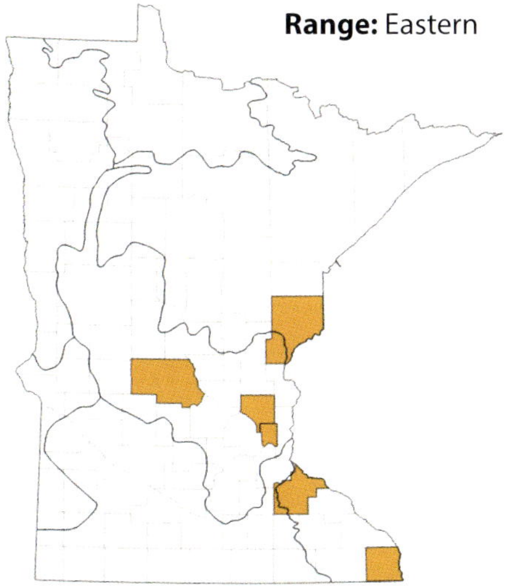

Field Notes: Associated with mature, hardwood forests.

Similar species: Unmistakable.

Flowers: No Minnesota floral records.

Flight Dates: May - June

Status: Rare. No modern records.

Common Wood Fly

The Common Wood Fly (***Blera badia***) is a medium-sized, Nearctic flower fly. The face is light-colored with a black medial stripe. The abdomen has orange or yellow markings on the edges of segments 2 and 3. The hind femur is mostly yellow and the apical three tarsomeres are black.

Field Notes: Associated with mature, hardwood forests.

Similar species: Most similar to *B. confusa*. It could also be confused with *Pipiza* species which perch with wings held open and which share the same habitat.

Flowers: Adults observed in the vicinity of wild raspberries in Minnesota.

Flight Dates: June

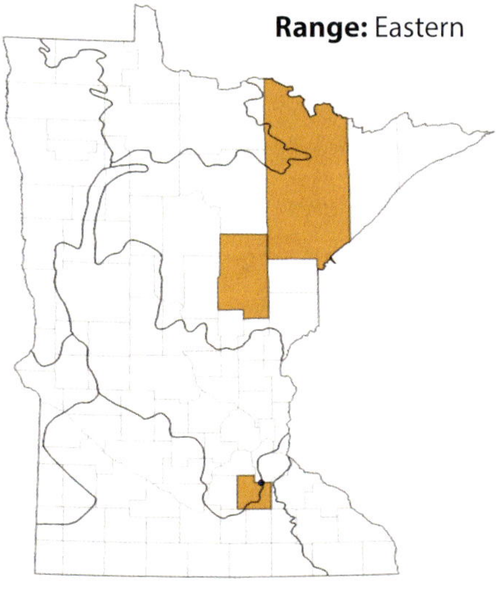

Range: Eastern

Status: Uncommon

Additional Minnesota Species Not Included in the Guide

Blera armillata: * Northern; face is orange; abdomen unmarked

Blera nigra: * Northern; legs and abdomen black; lower part of face is black

Confusing Wood Fly

The Confusing Wood Fly (***Blera confusa***) is a medium-sized, Nearctic flower fly. The face is light-colored with a black medial stripe. The abdomen has orange or yellow markings on the edges of segments 2 and 3. The hind femur is black and the apical two tarsomeres are black.

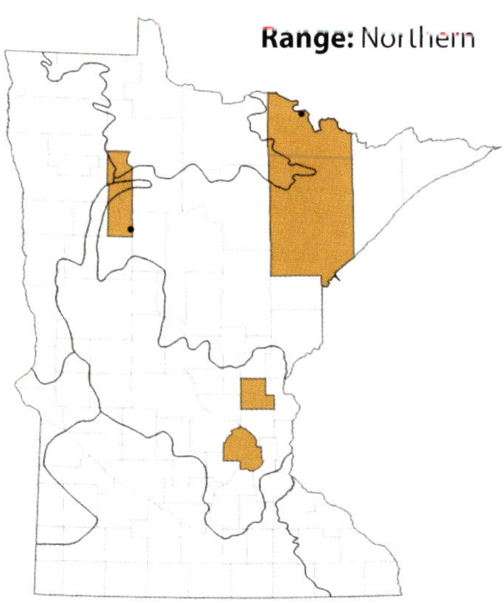

Field Notes: Many observations associated with bog habitat.

Similar species: Most similar to *B. badia*. It could also be confused with *Pipiza* species which perch with wings held open and which share the same habitat.

Flowers: Observed on bunchberry in Manitoba.

Flight Dates: May - June

Status: Rare

Subtribe Criorhina > Genus *Criorhina* :: Bumbleflies

The genus Criorhina includes large, bumblebee mimics. The R4+5 vein is straight. The crossvein r-m connects to cell dm at its middle. The face has a significant prominence to which the antennae attach. They have a long face, produced and extended downward with a pronounced bump. Larvae develop in rotting leaf litter.

Bare-cheeked Bumblefly

The Bare-cheeked Bumblefly (***Criorhina nigriventris***) is a large, Nearctic flower fly. It is a convincing bumblebee mimic. The lower side of the projected face is bare and black in color. The hind tibia is black. The larvae develop in decaying leaf litter.

Field Notes: This species flies very early in the season. Males are known to hilltop.

Similar species: Nearly identical to *C. verbosa*. But also similar to *Mallota* and other bumblebee mimics, wing venation differs.

Flowers: Adults visit wild plum and willow.

Flight Dates: April - May

Status: Rare

Range: Northern

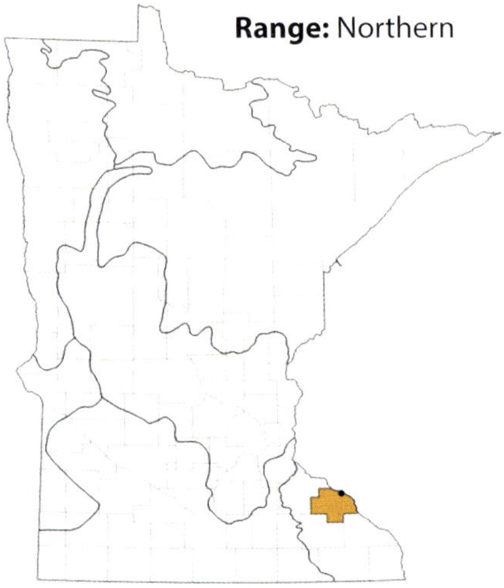

Additional Minnesota Species Not Included in the Guide

Criorhina verbosa: Eastern; the lower side of the projected face is pilose; hind tibia is orange with a black band in the middle.

Subtribe Criorhina > Genus *Sphecomyia* Yellowjacket Flies

The genus *Sphecomyia* includes large, very convincing social wasp mimics with either a very long antennae or, in many western species, antennae with shorter and rounder flagellum. Interestingly, the species with the short antennae have front feet that are black, as in *Temnostoma*. The biology of these flies is not well known. A single species is known from Minnesota

Long-horned Yellowjacket Fly

The Long-horned Yellowjacket Fly (***Sphecomyia vittata***) is a large, Nearctic flower fly. It has long antennae with the arista attached dorsally to the flagellum. The behavior and the bold yellow-and-black markings make this fly a most convincing mimic of yellowjacket hornets. The larvae are unknown but probably develop in rot holes like other flies in tribe Milesiini.

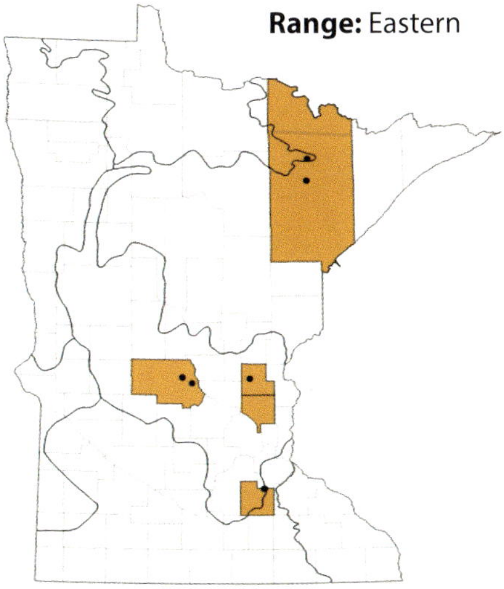

Range: Eastern

Status: Rare

Field Notes: Very little is known about the biology of these flies except that they seem to be associated with deciduous forests with old trees. From the few records in Minnesota, it appears to have a relatively short, early flight season.

Similar species: This convincing mimic is most likely to be mistaken for a hornet. The long antennae should set it apart from similar-colored eristalines.

Flowers: Mustard and mock orange in Minnesota.

Flight Dates: May - June

Subtribe Milesiina > Genus *Milesia* :: Giants

The flies of genus *Milesia* are very large and most are mimics of social wasps. Of the eighty or so worldwide species, three are known from the Nearctic and only one of those has been observed in Minnesota.

Virginia Giant

The Virginia Giant (***Milesia virginiensis***) is a very large, Nearctic flower fly. The thorax and abdomen pattern consists of a bold mosaic of yellow, orange and black. The adult flies are territorial and loud. Because of their behavior, persistent hovering and buzzing noise and harmlessness, they are sometimes referred to as "Good news bees." The larvae live in decaying deciduous wood and rot holes.

Range: Southern

Field Notes:
A specimen was collected at the fish hatchery located across the Minnesota River from St. Peter on July 22, 1922.

Similar species:
Unmistakable

Flowers:
No floral records for Minnesota.

Flight Dates: July

Status:
Rare, no modern records

Subtribe Milesiina > Genus *Spilomyia* :: Hornet Flies

The genus Spilomyia includes large, wasp mimics with strongly-patterned and spotted eyes. The front half of the wing is darkened to resemble the narrow, folded wings of the vespid wasps. Rear femurs have sharp, apical spurs. The related genus *Temnostoma* differs from *Spilomyia* by the lack of spurs on their rear legs and by unmarked eyes. Larvae develop in water-filled rot holes in deciduous trees. Four species are found in Minnesota.

Four-lined Hornet Fly

Four-lined Hornet Fly (***Spilomyia sayi***) is a large, mostly black, Nearctic species. A convincing mimic of potter wasps. Wings are darkened along the fore-edge, resembling the folded wings of vespid wasps. The abdomen curves downward and segments 2 and 3 have thin, yellow bands. Eyes are patterned with dots and blotches, and are darker than other species in the genus.

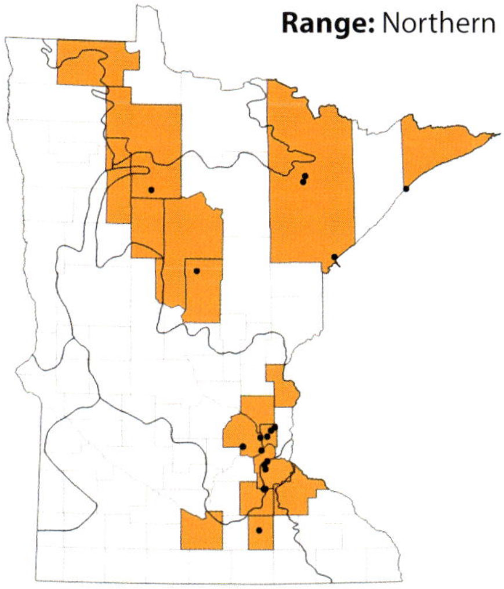

Range: Northern

Field Notes: Look for this species in prairie meadows adjacent to old woods.

Similar species: Most similar to *Temnostoma* and *Doros*.

Flowers: Adults are often found on goldenrod flowers, but have also been observed on black cohosh, and various asters.

Flight Dates: June - October

Status: Common

Broad-banded Hornet Fly

The Broad-banded Hornet Fly (***Spilomyia alcimus***) is a large, yellow-and-black Nearctic flower fly. A very convincing social wasp mimic. The abdomen is broadly banded in yellow in a manner very similar to that of queen yellowjackets (*Vespula* spp.), two wide, yellow bands on each segment. The face is yellow. The eyes are patterned. The scutellum is mostly yellow. The legs are yellow except for the front legs where the tarsi and tibia are black.

Field Notes: Adults seem strongly associated with Canada elderberry, a flower not utilized by many other insects.

Range: Eastern

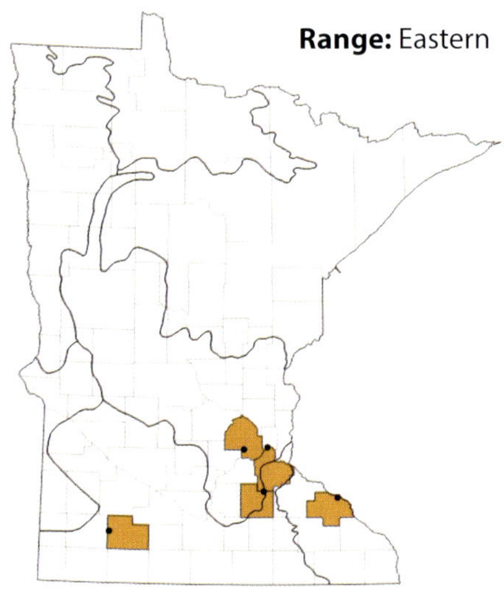

Similar species: Most similar to other yellow-and-black social wasp mimics such as *Somula decora, Sphecomyia vittata*, or *Temnostoma alternans*.

Flowers: Canada Elderberry and Gray Dogwood in Minnesota.

Flight Dates: June - July

Status: Uncommon

Bald-faced Hornet Fly

The Bald-faced Hornet Fly (***Spilomyia fusca***) is a large, black-and-white Nearctic flower fly. A convincing mimic of the aggressive social wasp of the same name, the Bald-faced Hornet. Segment 2 of the abdomen is entirely black. The face is white. The eyes are light-colored and patterned with black dots.

Range: Northern

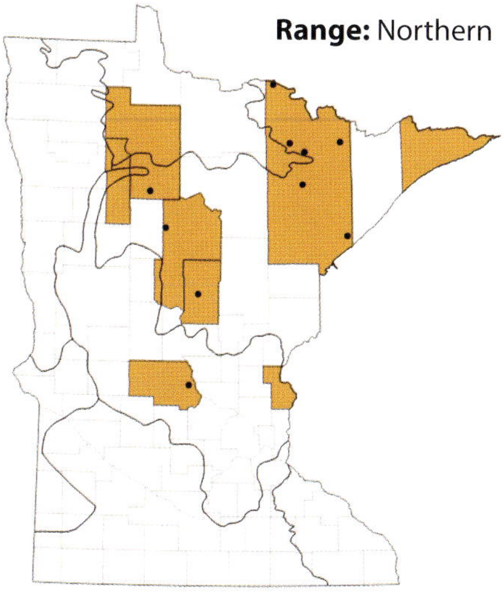

Field Notes: Most common in the mixed-boreal part of the state.

Similar species: Most similar to *Sericomyia militaris* and to its model the Bald-faced Hornet.

Flowers: In Minnesota, adults have been observed on tansy, goldenrod, coneflower, boneset, and joe-pye weed.

Flight Dates: July - September

Status: Common

Eastern Hornet Fly

The Eastern Hornet Fly (***Spilomyia longicornis***) is a large, black-and-yellow Nearctic flower fly. A convincing wasp mimic. The abdomen is black with thin yellow bands, two bands on each segment. In addition to being a visual mimic, the Eastern Hornet Fly mimics the behavior of wasps as well, wagging its abdomen up and down, flicking its wings, and waving its dark-colored front legs as if they were long antennae. The unusual eye coloration and patterning, dark spots and blotches overlaying a light background, is characteristic of this species and other species in this genus. Larvae are obligate rot-hole inhabitants.

Range: Eastern

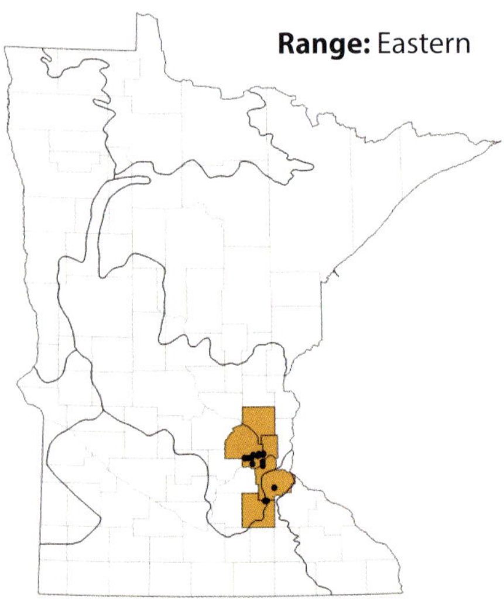

Status: Common

Field Notes: There exists a curious intersection here, between language and mimicry, as the species name "longicornis" refers to the Longhorn Beetles. My guess is that Loew, who assigned the name in 1872, had the Locust Borer (*Megacyllene robiniae*) in mind, a beetle that has a somewhat similar yellow-banding pattern. Expressed in terms of what it is not, the name draws attention to the convergent evolution of mimicry.

Similar species: Most similar to *Spilomyia sayi* but could also be mistaken for other yellowjacket mimics.

Flowers: In Minnesota, adults have been observed at asters, goldenrod, wild onion, wild cucumber, fleabane, joe-pye weed, and white snakeroot.

Flight Dates: August - October

Subtribe Temnostomina > Genus *Pterallastes* :: Goldenbacks

The genus Pterallastes is small, with only four species: three in Asia, one in the Nearctic. These are large, long-bodied flies. The larvae are undescribed.

Goldenback

The Goldenback (***Pterallastes thoracicus***) is a large, Nearctic flower fly. The R4+5 wing vein is sinuous. The hind femurs are black and relatively straight, not inflated. The thorax is covered in golden pile. The abdomen is black.

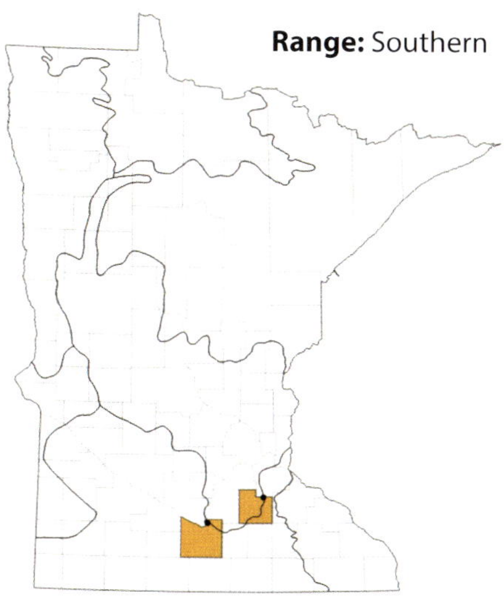

Range: Southern

Field Notes: A single photo record exists for Minnesota.

Similar species: Most similar to *Brachypalpus* and *Mallota*

Flowers: Mock-orange shrub and American bellflower in Minnesota.

Flight Dates: June

Status: Rare

Subtribe Temnostomina > Genus *Temnostoma* Falsehorns

The genus *Temnostoma* includes relatively large species that mimic hornets (Vespinae) and mason wasps (Eumeninae). The thorax is matte-black with yellow or light-gray pollinose markings and spots. The abdomen segments have one or two transverse bands that can be yellow or white. The front legs are black. When threatened or in flight, the flies extend their black forefeet in front of their heads and wave them, imitating the antennae of wasps. Species in the genus *Temnostoma* differs from the related genus *Spilomyia* by the lack of spurs on the rear legs and unmarked eyes. Falsehorns occur in moist woodland with mature stands of deciduous trees such as maple, birch, and aspen. Larvae feed and burrow in decaying wood.

The genus includes a total of twenty-three species, eight of which are found in North America. Five species are known from Minnesota.

Key to the Temnostoma species

1 Social wasp mimic. Abdomen banding consists of pairs of yellow or white bands 2
1′ Mason wasp mimic. Abdomen banding consists of a single yellow or white band along the anterior edge of each segment 4

2 Black abdomen with yellow bands 3
2′ Black abdomen with pairs of thin, white bands .... ***T. venustum***

3 Yellow bands meet medially and enclose a pair of black rectangles on the 3rd and 4th abdomen segments .... ***T. excentrica***
3′ Well-defined yellow triangle on thorax above scutellum .... ***T. alternans***

4 Wings with several basal cells clear, free from hair 5
4′ Wings without clear cells, completely microtrichose .... ***T. trifaciatum***

5 Pair of faint, pollinose spots on thorax above scutellum. Bare metasternum. Black setae on black portion of final abdominal segment. Wing cells clear or partly clear .... ***T. barberi***
5′ Yellow setae on black portion of final abdominal segment. Metasternum with hair. Without pollinose spots on thorax above scutellum .... ***T. balyras***

Wasp-like Falsehorn

The Wasp-like Falsehorn (***Temnostoma alternans***) is a large, yellow-and-black, Nearctic flower fly. A convincing yellowjacket wasp (Vespinae) mimic. In addition to being a visual mimic, this fly (and the other Falsehorns) lift and wave their black forefeet in front of their heads, imitating the antennae of wasps. Markings on the abdomen consist of two yellow bands on each segment beginning with the 2nd segment. On segment 2 these bands do not touch in the middle and are unequal in length and thickness. As this pattern continues, the bands thicken and connect in the middle, leaving a central black diamond and two lateral black spots on segment 4. The larvae are found in decaying wood.

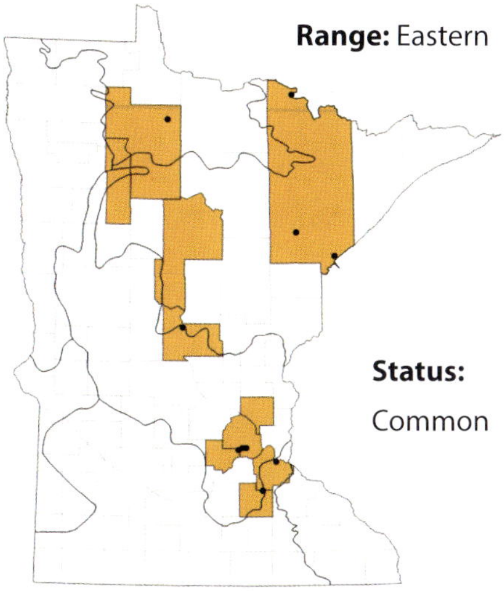

Range: Eastern

Status: Common

Field Notes: As shown in the photo, females and males can be quite different in size.

Similar species: Similar to *Temnostoma excentrica, Spilomyia alcimus, Somula decora*, and *Sphecomyia vittata*, which are all yellowjacket mimics.

Flowers: In Minnesota, adults have been observed on common milkweed, apple, wild rose, and elderberry.

Flight Dates: June - July

Yellow-haired Falsehorn

The Yellow-haired Falsehorn (***Temnostoma balyras***) is a black-and-yellow, Nearctic flower fly. A convincing mason wasp (Eumeninae) mimic. Markings on the thorax are yellow and pollinose (pollen-like). Abdomen markings consist of a single yellow band at the front of each segment — 3 bands are visible (segments 2-4) on males and 4 bands on females. The male's final abdomen segment is yellow pilose. Wing cells bm and cua only partially bare. The metasternum is pilose.

Range: Eastern

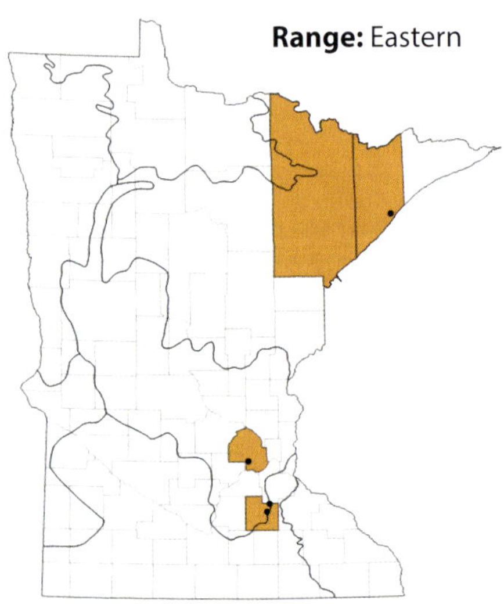

Field Notes: A known hilltopper.

Similar species: Nearly identical to *Temnostoma barberi, Temnostoma trifasciatum,* and *Temnostoma obscurum*. Care must be taken to separate these species.

Flowers: Dogwood is the only floral record in Minnesota so far.

Flight Dates: May - June

Status: Common

Bare-bellied Falsehorn

The Bare-bellied Falsehorn (***Temnostoma barberi***) is a black-and-yellow, Nearctic flower fly. A convincing mason wasp (Eumeninae) mimic. Markings on the thorax are yellow and pollinose (pollen-like). Abdomen markings consist of a single yellow band at the front of each segment — 3 bands are visible (segments 2-4) on males and 4 bands on females. A pair of faint pollinose spots above the scutellum (useful for photo identification). Wing cells bm and cua are extensively bare. The metasternum is bare.

Range: Eastern

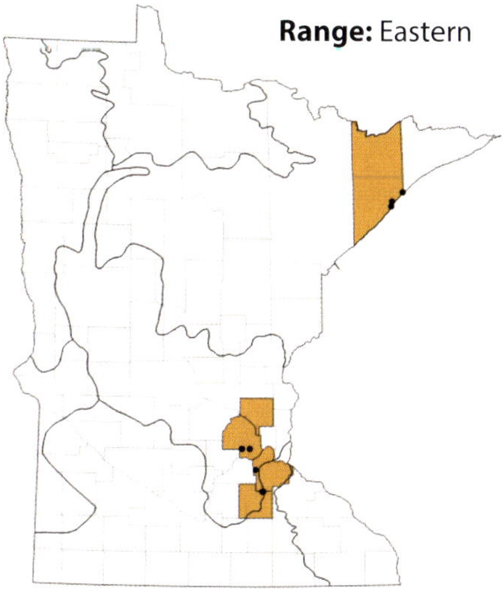

Status: Common

Field Notes: Can have white banding in the north.

Similar species: Nearly identical to *Temnostoma balyras, Temnostoma trifasciatum*, and *Temnostoma obscurum*. Care must be taken to separate these species.

Flowers: Dogwood, sumac, wild rose, and brambles in Minnesota.

Flight Dates: June - July

Black-spotted Falsehorn

The Black-spotted Falsehorn (***Temnostoma excentrica***) is a large, yellow-and-black, Nearctic flower fly. A convincing Yellowjacket Wasp (Vespinae) mimic. In addition to being a visual mimic, this fly (and the other Falsehorns) lift and wave their black forefeet in front of their heads, imitating the antennae of wasps. Markings on the thorax are whitish-gray in color. Paired yellow bands are joined medially and enclose a pair of black, rectangular spots on the 3rd and 4th abdomen segments (not joined on segment 2). In the north, the abdominal markings can be white instead of yellow, mimicking the bald-faced hornet (*Dolichovespula maculata*). The larvae are found in decaying wood.

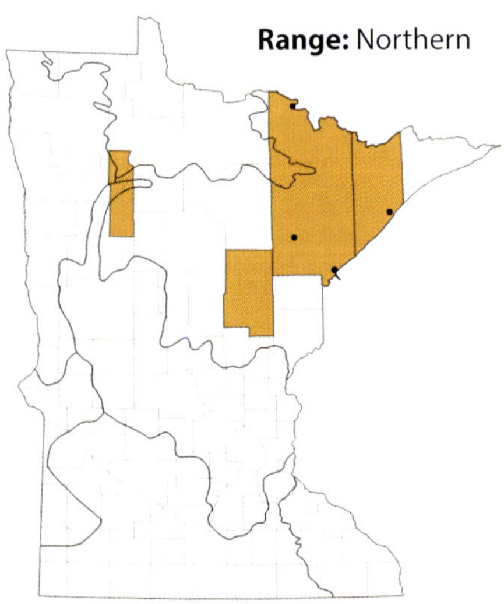

Range: Northern

Status: Uncommon

Field Notes: Most likely in mixed-boreal habitats in Minnesota.

Similar species: Most similar to *Temnostoma venustum* but could also be confused with *Temnostoma alternans, Spilomyia alcimus, Spilomyia fusca, Somula decora,* and *Sphecomyia vittata*, which are all yellowjacket mimics.

Flowers: No floral records for Minnesota. Observed on elderberry in Ontario.

Flight Dates: June - July

Black-banded Falsehorn

The Black-banded Falsehorn (***Temnostoma venustum***) is a large, white-and-black, Nearctic flower fly. A convincing baldface hornet mimic. In addition to being a visual mimic, this fly (and the other Falsehorns) lift and wave their black forefeet in front of their heads, imitating the antennae of wasps. The while mark in front of the wing on the top of the thorax is divided into two spots. Paired white bands that do not meet medially on each abdomen segment. The bands on segment 2 are often faded. The larvae are found in decaying wood.

Range: Eastern

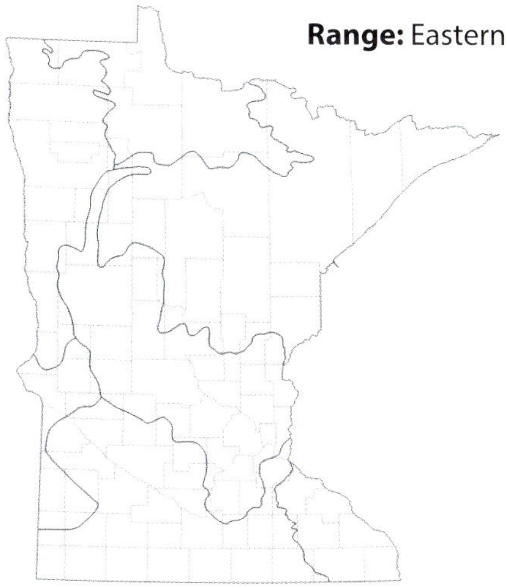

Status: Hypothetical

Field Notes: Most likely to be found in mixed-boreal habitats in Minnesota.

Similar species: Most similar to *Temnostoma excentrica* but could also be confused with *Spilomyia fusca*, another bald-faced hornet mimic.

Flowers: No floral records for Minnesota. Elsewhere it has been observed visiting dogwood and smartweed.

Flight Dates: July

Additional Minnesota Species Not Included in the Guide

Temnostoma trifasciatum: * Southern; wing cells bm and cua are bare

Temnostoma obscurum: Northern; wing cells bm and cua completely microtrichose; white bands

Subtribe Temnostomina > Genus *Teuchocnemis* Spur Flies

The genus *Teuchocnemis* is small, with only two species, both in the Nearctic. The R4+5 vein is slightly sinuous. Marginal cell r1 is open. CuA+CuP is noticeably extended and doesn't reach the wing margin. The hind tibia has a spur. One of the two species has been observed in Minnesota.

Black Spur Fly

The Black Spur Fly ***(Teuchocnemis lituratus***) is a large, Nearctic flower fly. Pollinose stripes on thorax like those of *Tropidia*. The front feet are black like those of *Temnostoma*. The antennae and face are yellowish-orange. The scutellum is orange.

Range: Eastern

Field Notes: This species has only recently been observed in Minnesota for the first time. It is associated with rich, deciduous woods and fens.

Similar species: Most similar to *Blera*.

Flowers: No floral records for Minnesota.

Flight Dates: April - June

Status: Rare

Subtribe Tropidiina > Genus *Syritta* :: Compost Flies

The genus *Syritta* includes small, elongated species with strikingly swollen hind femurs. The larvae develop in decaying organic matter. The genus is widespread with at least 60 species worldwide, the greatest diversity occurring in Africa. Only two species occur in North America, both adventive, one introduced from Europe, the other from Africa.

Common Compost Fly

The Common Compost Fly (***Syritta pipiens***) is a small, Palearctic flower fly. It was introduced to North America in the late 1800s. The wildly-inflated hind femur is unique to this species. The lower part of the face protrudes as it does for flies of the subtribe Spheginina. Also, the pale coloration on the side of the thorax and the yellow, triangular markings on the second and third abdominal segments, and polinose spots on segment 4 are helpful for identification. The larvae develop in compost, manure, and decaying organic matter.

Range: Southern

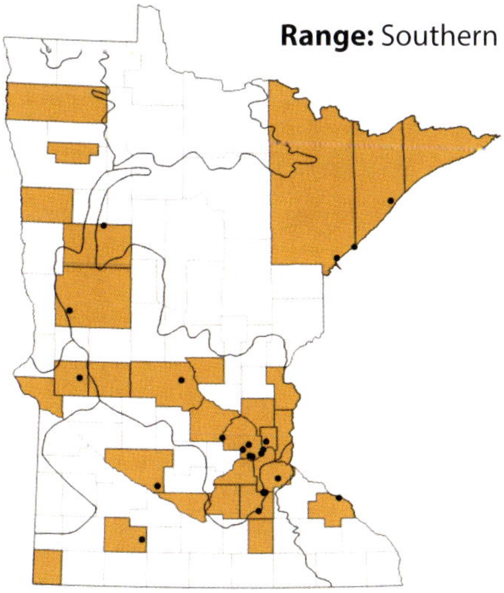

Status: Abundant

Flight Dates: June - October (most observations in September)

Field Notes: The males, though small, are quite pugnacious and can often be observed flying and defending territories on or near garden flowers.

Similar species: Most similar to *Syritta flaviventris*, a more southerly species which has a large projection at the base of the hind femur. It could also be mistaken for small species of *Xylota*.

Flowers: White snakeroot, smartweed, aster, goldenrod, yellow wood sorrel, mountain mint, devil's beggartick, and cinquefoil in Minnesota.

Subtribe Tropidiina > Genus *Tropidia* :: Thicklegs

The genus *Tropidia* includes medium-sized species with tapered abdomens and enlarged hind femurs with a ventral flange. The R4+5 vein is slightly sinuous and the marginal cell r1 is open. One species is known from Minnesota while two additional species are definitely possible though their preferred habitats have been greatly reduced in modern times.

Common Thickleg

The Common Thickleg (***Tropidia quadrata***) is a medium-sized, Nearctic flower fly. It has a white polinose face with a medial ridge, black medial stripe and a black spot on its cheek. The hind femur is enlarged with a ventral flange and is entirely black. Two wide, polinose stripes on the thorax fade toward the rear of the thorax. The abdominal segments 2 and 3 have pairs of yellow marks (larger in the males). The end of the abdomen curves downward. The larvae are associated with water and decaying organic matter.

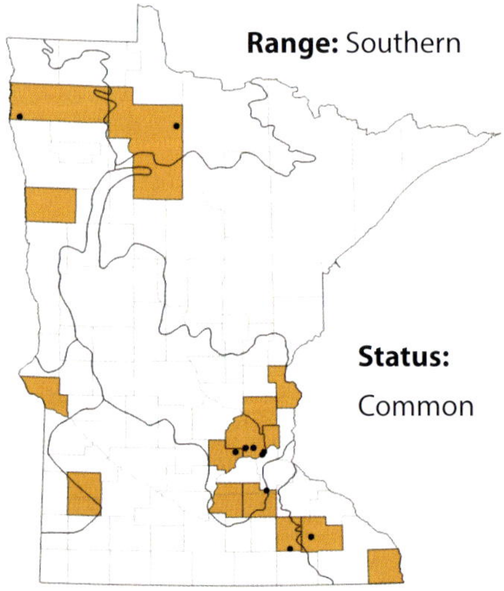

Range: Southern

Status: Common

Field Notes: The Common Thickleg tends to fly instead of walking (unlike the similar *Xylota* species).

Similar species: Most similar to other *Tropidia* and *Syritta* but could be mistaken for *Xylota*.

Flowers: Meadow anemone, iris, common milkweed, swamp milkweed, and purple prairie clover in Minnesota.

Flight Dates: June - August

Additional Minnesota Species Not Included in the Guide

Tropidia calcarata: * Eastern; a water lily specialist

Tropidia mamillata: * Southern; a tallgrass prairie specialist

Subtribe Xylotina > Genus *Brachypalpus* :: Catkin Flies

The flies of genus *Brachypalpus* are large, long-bodied flower flies. They have triangular-shaped faces. The R4+5 vein is straight. One species is known from Minnesota with a second species possible.

Eastern Catkin Fly

The Eastern Catkin Fly (***Brachypalpus oarus***) is a large, Nearctic flower fly. An imperfect mimic of large mining bees. The body is covered with long, yellowish hair. The face of the female is triangular in shape (rounded in *Chalcosyrphus*). All black hind tibia. The larvae live in rot holes in deciduous trees.

Field Notes: This infrequently observed fly is associated with old woodlands where it is usually observed hovering near to or perched on sunlit tree trunks. According to Telford, ten males and one female were reared from pupae found in a rotten tree stump with a large colony of carpenter ants found along Coon Creek near Saint Paul in 1936.

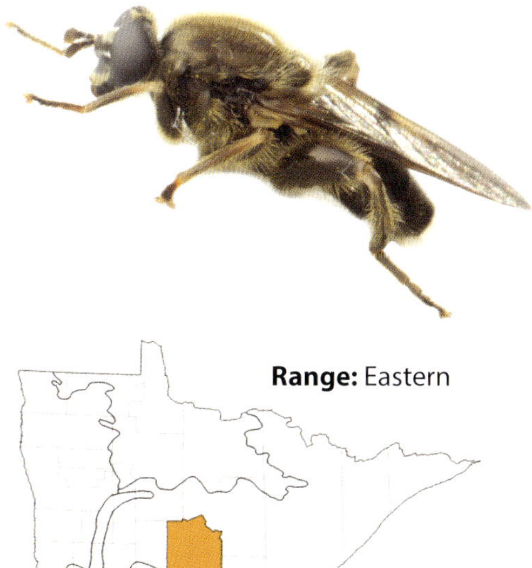

Range: Eastern

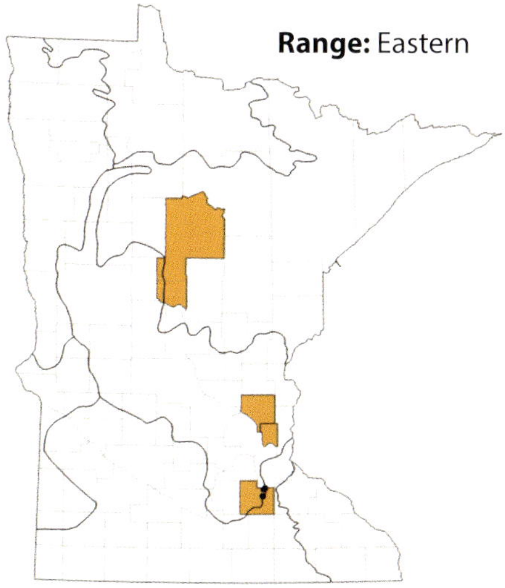

Status: Uncommon

Similar species: Most similar to *Brachypalpus femoratus* and *Chalcosyrphus inarmatus* (has orange at the base of hind tibia). Because of its behavior, it could also be mistaken for a robber fly of the genus *Laphria*.

Flowers: Pasque flower in Minnesota but it probably visits tree flowers in the canopy and catkins as well.

Flight Dates: April - June

Additional Minnesota Brachypalpus Species Not Included in the Guide

Brachypalpus femoratus: * Western; black face; orange pile on thorax.

Subtribe Xylotina > Genus *Chalcosyrphus*
Leafwalkers

The flies of the genus *Chalcosyrphus* are medium to large-sized, long-bodied flower flies. Rather convincing spider wasp and sawfly mimics. Several species have bright orange hind legs. The hind femurs are swollen. The face is concave and lacks a middle protuberance. The metasternum has long pile (very short in the closely related genus *Xylota*). The vein R4+5 is straight and the crossvein r-m meets cell dm at or beyond the middle. Larvae live in decaying sap under the bark of fallen trees.

Key to the Chalcosyrphus species

1 Entirely black; wings dark and iridescent .... ***C. chalybeus***
1′ Body or legs not entirely black 2

2 Abdomen segments 2-4 are red 3
2′ Abdomen segments are not red 4

3 Basal tarsomeres of rear leg are yellow .... ***C. libo***
3′ Basal tarsomeres of rear leg are black .... ***C. pigra***

4 Rear femur is bright orange 5
4′ Rear femur is black 6

5 Rear coxa is orange .... ***C. vecors***
5′ Rear coxa is black .... ***C. curvaria***

6 Abdomen is unmarked and black .... ***C. inarmatus***
6′ Abdomen is marked with yellow or muted-yellow spots 7

7 Rear tibia is black with a pale area at the basal joint .... ***C. nemorum***
7′ Base of rear tibia is yellow .... ***C. anthreas***

Dusky-banded Leafwalker

The Dusky-banded Leafwalker (***Chalcosyrphus nemorum***) is a small, Holarctic flower fly. The paired marks on abdominal segments 2 and 3 are muted-yellow in color and can appear silver or white. The hind femur is black and expanded toward the apex. The hind tibia is entirely black with a pale area at the basal joint.

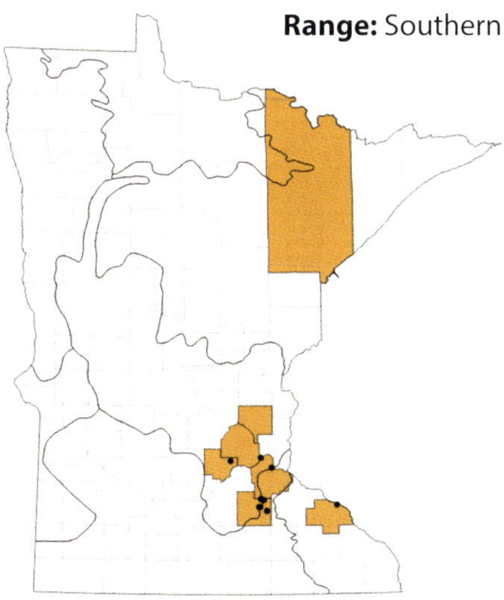

Range: Southern

Field Notes: The pupa pictured was found in the spring under bark on a fallen log near the edge of a fen and reared. Adults like to perch on sunlit wood, usually very close to the ground.

Similar species: Most similar to small *Xylota* species.

Flowers: Yellow rocket in Minnesota.

Flight Dates: May - September

Status: Common

Violet Leafwalker

The Violet Leafwalker (***Chalcosyrphus chalybeus***) is a large, Nearctic flower fly. A convincing spider wasp mimic. Entirely black with a metallic purple sheen. The wings are dark and iridescent blue.

Range: Eastern

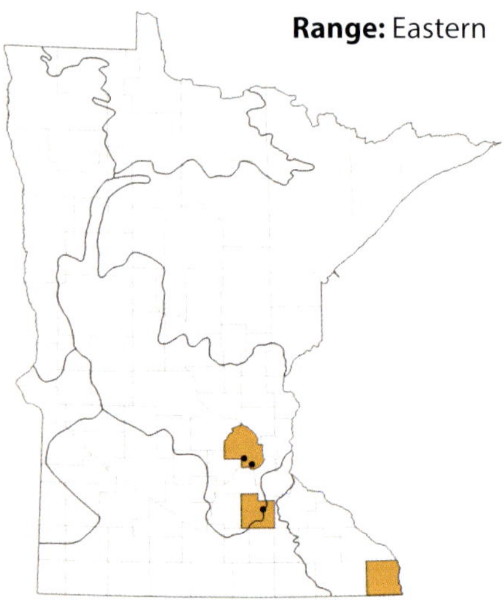

Status: Common

Field Notes: Look for this impressive fly in low vegetation where it behaves like a spider wasp, rarely resting, flicking its wings.

Similar species: Most similar to the wasps it mimics. It shouldn't be confused with other flower flies.

Flowers: Virginia strawberry in Minnesota.

Flight Dates: June - July

Yellow-haltered Leafwalker

The Yellow-haltered Leafwalker (***Chalcosyrphus curvaria***) is a large, Nearctic flower fly. The front and middle legs are orange. The rear femur is bright orange with just the apex black. The hind coxa are black. The halter is also yellow or orange.

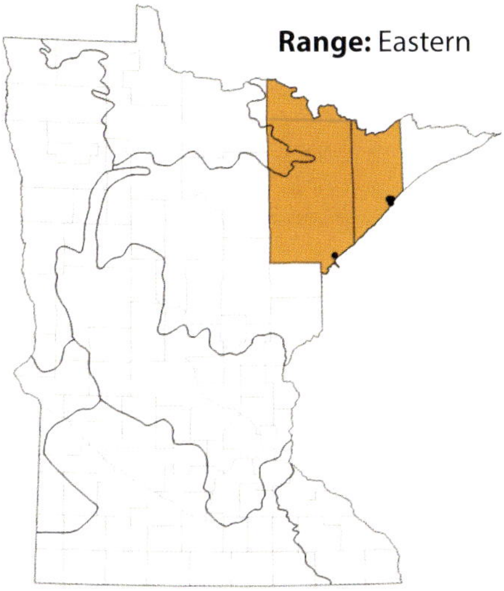

Range: Eastern

Field Notes: Found predominantly in mixed-boreal habitats.

Similar species: Most similar to *C. vecors* which has the hind coxa orange.

Flowers: No floral records for Minnesota.

Flight Dates: July - August

Status: Uncommon

Long-haired Leafwalker

The Long-haired Leafwalker (***Chalcosyrphus libo***) is a medium-sized, Nearctic flower fly. The abdomen segments 2 - 4 are a deep reddish-orange with a black, medial mark at the beginning of segment 2. The pile on the thorax is erect. The basal tarsomeres of the rear leg are yellow.

Field Notes: Single photo observation is from a sunlit tree trunk in northern Minnesota.

Similar species: Most similar to *C. piger* which has black tarsomeres on its rear leg.

Flowers: No floral records for Minnesota.

Flight Dates: June

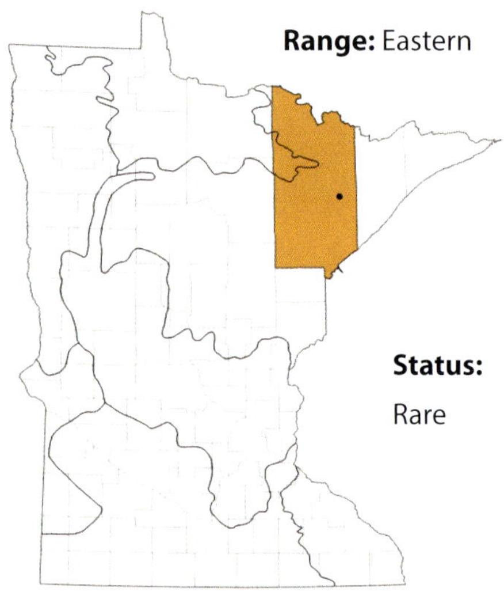

Range: Eastern

Status: Rare

Additional Minnesota Species Not Included in the Guide

Chalcosyrphus anthreas: * Eastern; similar to *C. nemorum*

Chalcosyrphus inarmatus: Northern; similar to *Brachypalpus oarus*; rear tibia partly orange

Chalcosyrphus nigromaculatus: * Western; similar to *C. metallifer*

Chalcosyrphus piger: Southern; abdomen segments 2-4 red; basal tarsomeres of rear leg black

Chalcosyrphus vecors: Northern; orange femur; orange coxa

Subtribe Xylotina > Genus *Xylota* :: Leafwalkers

The flies of the genus *Xylota* are small to medium-sized, long-bodied flower flies. Rather convincing spider wasp and sawfly mimics. They are rarely observed on flowers. Instead, they are most often observed perched and walking on leaves where they feed upon honeydew and fallen pieces of pollen. The metasternum appears velvety, with very short pile (very long pile in the closely related genus *Chalcosyrphus*). The vein R4+5 is straight and the crossvein r-m meets cell dm at or beyond the middle. The larvae live in decaying timber and other woody debris. Note: these flies tend to look exactly alike in photographs and often require careful examination of a specimen with a microscope to secure an identification.

Key to the male Xylota species

1 Abdominal segments 2 and 3 entirely orange .... ***X. flavitibia***
1′ Abdominal segments 2 and 3 not entirely orange 2

2 Abdominal segments 2 and 3 black with dark, shiny spots, diffusely white or yellow 3
2′ Abdominal segments 2 and/or 3 with yellow-orange spots 5

3 Anepisternum pale pilose .... ***X. flavifrons***
3′ Anepisternum partly black pilose 4

4 Long spine on metatrochanter .... ***X. naknek***
4′ Short, indistinct spine on metatrochanter .... ***X. tuberculata***

5 A pair of small, oval spots on segment 2, with segment three entirely black .... ***X. angustiventris***
5′ Spots on both abdominal segments 2 and 3 6

6 Katepisternum shiny on ventral 2/3 .... ***X. quadrimaculata***
6′ Katepisternum not shiny 7

7 Spine on metatrochanter longer than the width of the spine's base .... ***X. annulifera***
7′ Spine on metatrochanter absent or indistinct 8

8 Segments 2 and 3 with large yellow-orange spots. Spots with thin, black margins .... ***X. subfasciata***
8′ Segments 2 and 3 with yellow-orange spots surrounded by thick, black margins 9

9 Frontal triangle extensively bare and shiny .... ***X. confusa***
9′ Frontal triangle mostly polinose, with small, bare area above antennae .... ***X. hinei***

Yellow-footed Leafwalker

The Yellow-footed Leafwalker (***Xylota flavitibia***) is a medium-sized Nearctic flower fly. Segments 2 and 3 are entirely orange with no black, medial mark on segment 2. The basal tarsomeres of the rear leg are yellow.

Range: Northern

Status: Rare

Field Notes: One specimen from Cook County collected in 1973.

Similar species: *Chalcosyrphus libo*

Flowers: No floral records for Minnesota but known to visit *Fragaria* and *Heracleum* elsewhere.

Flight Dates: July

Two-spotted Leafwalker

The Two-spotted Leafwalker (***Xylota angustiventris***) is a medium-sized Nearctic flower fly. The male has two, ivory spots on segment 2 of the abdomen. The abdomen of the female is entirely black with no marks. The alula on the wing is bare in the middle (completely trichose in other species).

Range: Southern

Field Notes: No modern records. A few bedraggled museum specimens exist from southeastern Minnesota.

Similar species: Easily confused with other *Xylota.*

Flowers: No floral records for Minnesota.

Flight Dates: May - June

Status: Rare

Large-spotted Leafwalker

The Large-spotted Leafwalker (***Xylota subfasciata***) is a medium-sized, Nearctic flower fly. Segments 2 and 3 have large yellow spots; for males, the segments are almost entirely yellow. The apical two tarsomeres on all the legs are black.

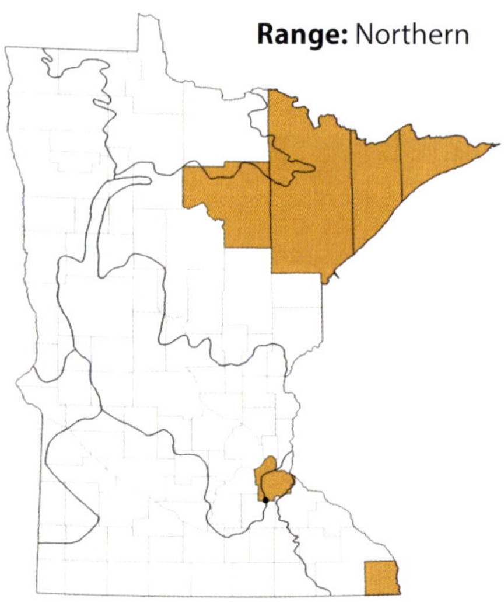

Range: Northern

Status: Uncommon

Field Notes: Associated with hardwood forests.

Similar species: Most similar to *X. annulifera.*

Flowers: Observed on leaves beneath wild geranium in Minnesota.

Flight Dates: May - June

Four-spotted Leafwalker

The Four-spotted Leafwalker (***Xylota quadrimaculata***) is a medium-sized, Nearctic flower fly. Abdomen segments 2 and 3 have yellow spots. The spots on segment 2 of the males are trapezoid-shaped. Female spots are reduced in size compared to males. The katepisternum is shiny like obsidian ventrally, approximately the bottom 2/3.

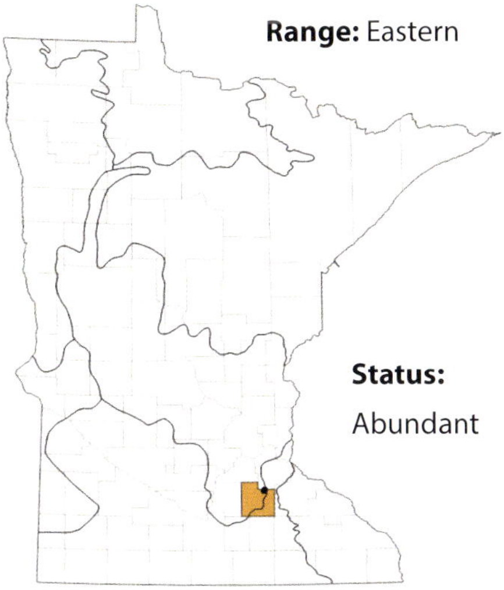

Range: Eastern

Status: Abundant

Field Notes: Often found perching on bare ground in forest openings and edges.

Similar species: Most similar to *X. annulifera.*

Flowers:
No floral records for Minnesota.

Flight Dates: June - September

Hine's Leafwalker

The Hine's Leafwalker (***Xylota hinei***) is a small, Nearctic flower fly. Segments 2 and 3 have yellow spots surrounded by thick black margins. Males have a curved, bare area above the antennae on the frons. Females have black, bristles on the postalar callus (requires microscope).

Range: Northern

Field Notes: Known from a single, modern record.

Similar species: Males are most similar to *X. confusa* males.

Flowers: No floral records for Minnesota but observed feeding on burdock pollen from leaves below flowers.

Flight Dates: June

Status: Rare

Longspine Leafwalker

The Longspine Leafwalker (***Xylota annulifera***) is a medium-sized Nearctic flower fly. Segments 2 and 3 have yellow spots. The three, apical tarsomeres are black. The front and middle tibia are black on the apical end. The spines on rear trochanters are longer than the width of the spine's base.

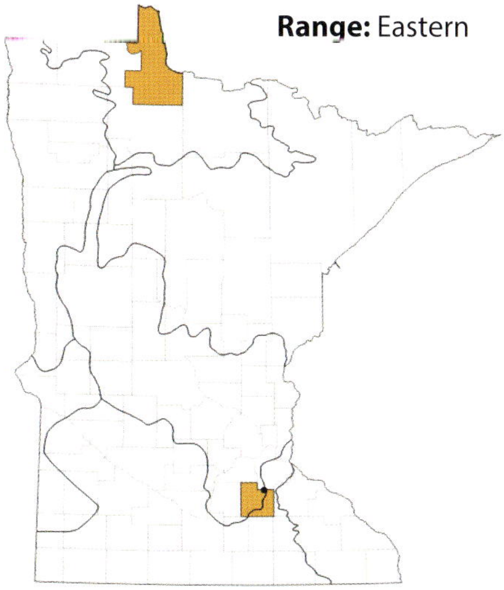

Range: Eastern

Field Notes: Known from deciduous and mixed-boreal habitats.

Similar species: Most similar to *X. subfasciata.*

Flowers: No floral records for Minnesota.

Flight Dates: June

Status: Uncommon

Confusing Leafwalker

The Confusing Leafwalker (***Xylota confusa* = *Xylota bigelowi***) is a medium-sized Nearctic flower fly. The two, apical tarsomeres are black. The front and middle tibia are black on the apical end. Segments 2 and 3 of males have yellow spots. Female with metallic blue paired spots, black and polinose.

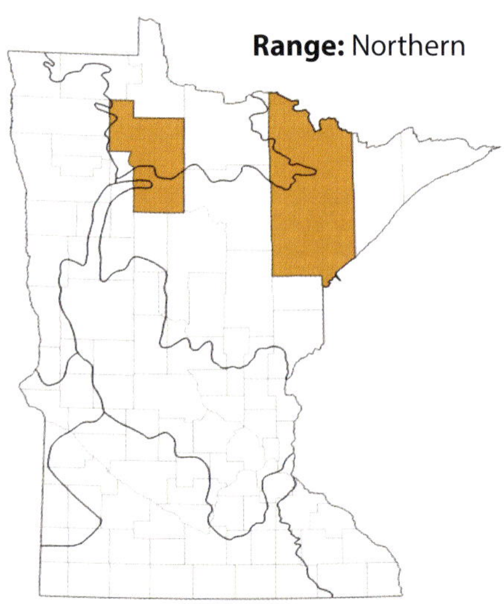

Range: Northern

Field Notes: Known from a single, modern record at Itasca State Park.

Similar species: Males are most similar to *X. hinei*. Because of the difference in coloration, females are most similar to *X. flavifrons, X. tuberculata, and X. naknek.*

Flowers: No floral records for Minnesota.

Flight Dates: June

Status: Uncommon

Naknek Leafwalker

The Naknek Leafwalker (***Xylota naknek***) is a medium-sized, Nearctic flower fly. Male and female with metallic blue paired spots that are polinose, diffusely white on a black abdomen. There is a long spine on the rear trochanter.

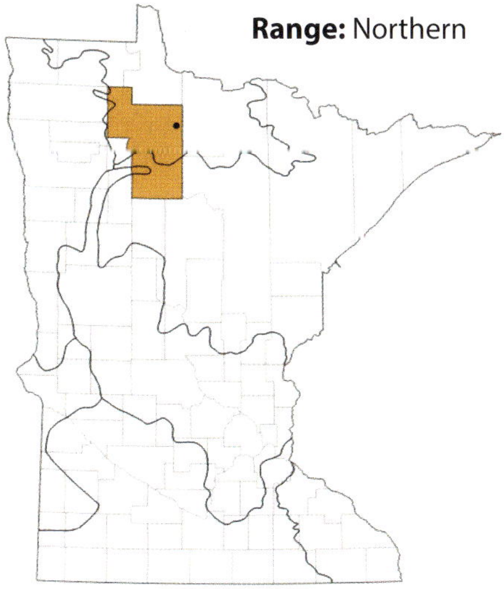

Range: Northern

Field Notes: Found in mixed-boreal habitat in Minnesota.

Similar species: Most similar to *X. flavifrons*, *X. tuberculata*, and female *X. confusa*.

Flowers: Sumac is the only record so far in Minnesota.

Flight Dates: June - July

Status: Uncommon

Additional Minnesota Species Not Included in the Guide

Xylota flavifrons: * Northern; similar to *X. naknek* but without spine on trochanter.

Xylota tuberculata: Northern; similar to *X. naknek* but with short spine on trochanter.

Tribe Rhingiini

Subtribe Cheilosiina > Genus *Cheilosia* :: Blacklets

Worldwide, one of the most speciose genera of flower flies. Also, perhaps, the most difficult. The larvae are phytophagous. Most species show strong host specificity, unfortunately next to nothing is known about the life histories of the species found in Minnesota. The only good news is that the genus is easily recognized because of their stout, black bodies and distinct facial tubercle. If you need to get to the truth of these flies you'll need to collect specimens and consult Skevington et. al. (2019) which covers many more species and the most recent keys by Fluke and Hull (1945, 1946) and Hull and Fluke (1950).

Inky Blacklet

The Inky Blacklet (***Cheilosia rita***) is small, Nearctic flower fly. The eyes are bare. The face is black but extensively dusted and long hairs are present on the lower part of the face. This pilosity on the face combined with bare eyes is unique among the Minnesota blacklet species. The larvae are phytophagous. The host plant is unknown.

Range: Northern

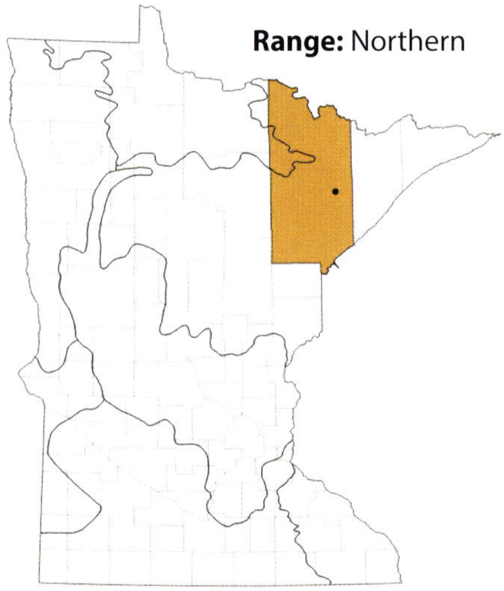

Status: Uncommon

Flowers: No floral records in Minnesota. Elsewhere it's been on flowering shrubs including wild plum and viburnum.

Flight Dates: May - June

Field Notes: More common in the mixed-boreal habitats of northern Minnesota.

Similar species: Easy to confuse with other blacklet species.

Black-backed Blacklet

The Black-backed Blacklet (***Cheilosia orilliaensis***) is a small, Nearctic flower fly. The scutellum rimmed with black bristles. The eyes are pilose. The face is bare and the lower portion is not produced forward (as in *C. hunteri*). The legs are partly yellow. The female is pale pilose, giving it a white look. The larvae are phytophagous. The host plant is unknown.

Range: Northern

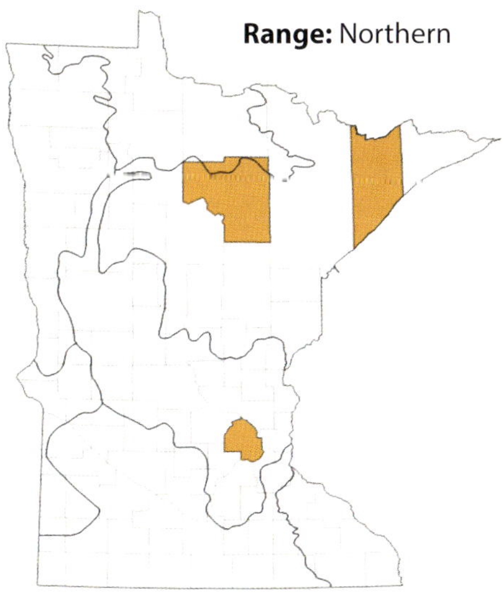

Status: Uncommon

Field Notes: Described as a new species by Horace Telford, *Cheilosia angelica*, and named for the flower it was found on, but is now recognized as a synonym of *C. orillianensis.*

Similar species: Most similar to *C. hunteri* but could be confused with other blacklets.

Flowers: Angelica in Minnesota.

Flight Dates: July

Steely Blacklet

The Steely Blacklet (***Cheilosia hunteri***) is a small, Nearctic flower fly. The eyes are bare. The face is not pilose. The antennal pits are joined. The scutellum is rimmed with black spines. The larvae are phytophagous. The host plant is unknown.

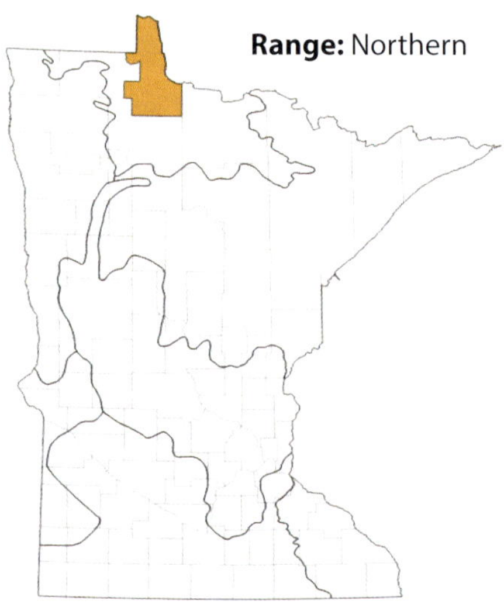

Range: Northern

Status: Uncommon

Field Notes: Most often found in mixed conifer forest habitats and is also associated with bogs.

Similar species: Most similar to *C. shannoni* but could be confused with other blacklets.

Flowers: No floral records in Minnesota. Elsewhere it has been observed visiting a number of flowering plants including joe-pye weed and meadowsweet.

Flight Dates: May - June

Jack Pine Blacklet

The Jack Pine Blacklet (***Cheilosia comosa***) is a small, Nearctic flower fly. Eyes are bare. The antennal pits are joined. The scutellum is plain, without spines. The female rear tibia is black and the basal tarsi are brown and do not appear much different than the rest of the tarsi. The larvae are phytophagous. The host plant is unknown.

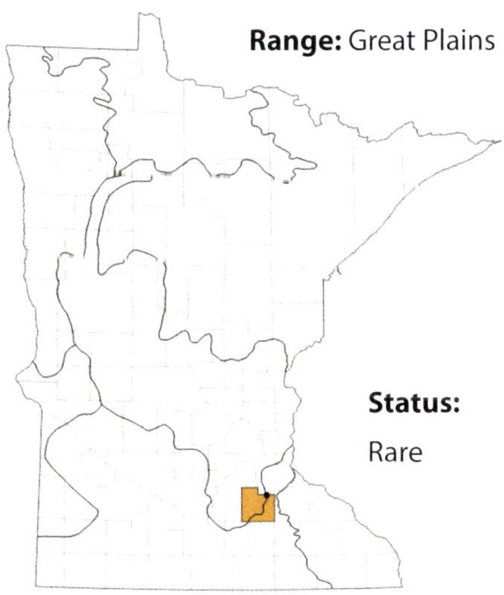

Range: Great Plains

Status: Rare

Field Notes: Thought to be associated with jack pine habitat, but has been found in sandy river bottom forest in Minnesota.

Similar species: Most similar to *C. prima* but could be confused with other blacklets.

Flowers: Dandelion is the only floral record for Minnesota.

Flight Dates: May - June

Additional Minnesota Species Not Included in the Guide

Cheilosia cottrelli: * Western
Cheilosia laterans: Northern
Cheilosia leucoparea: * Eastern
Cheilosia cynoprosopa: Southern
Cheilosia lasiophthalmus: * Northern
Cheilosia shannoni: * Eastern

Genus *Ferdinandea* :: Copperbacks

The flies of the genus Ferdinandea are recognized by the stipes on the top of the thorax and the curved, black bristles on the thorax just in front of the wings. Similar bristles rim the scutellum. The crossveins of the wings are darkened giving them a spotted look. One of the three Nearctic species is known from Minnesota.

Common Copperback

The Common Copperback (***Ferdinandea buccata***) is a small, Nearctic flower fly. The body is metallic bronze in color and covered with coppery pile. The top of the thorax is striped and there are a few curved, black bristles in front of the wings. Similar bristles rim the scutellum. The crossveins of the wings are darkened, giving them a spotted look. Arista is mostly yellow. The larvae are undescribed.

Range: Northern, but extending far to the south

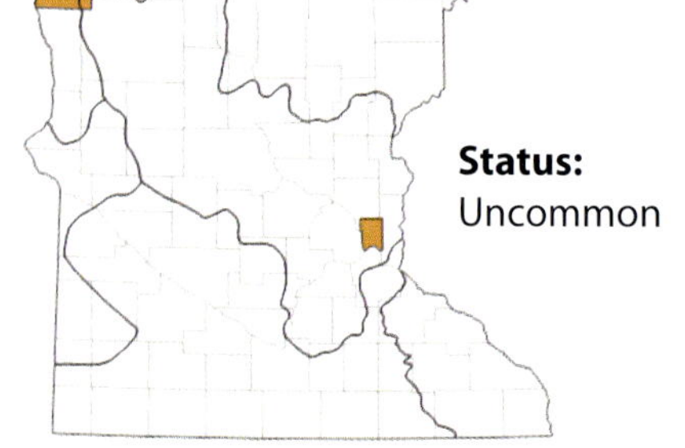

Status: Uncommon

Field Notes: Adults have been found in a wide variety of habitats.

Similar species: Most similiar to *F. croesus*, a western species with black arista.

Flowers: No floral records for Minnesota.

Flight Dates: May - June

Subtribe Rhingiina > Genus *Rhingia* :: Snout Flies

Only a single species from the genus *Rhingia* is found in North America. The flies of this genus are morphologically unique and easily recognized. The lower part of the face projects forward so far as to resemble a beak.

American Snout Fly

The American Snout Fly (***Rhingia nasica***) is a medium-sized Nearctic flower fly. The lower part of the face is preposterously protruded, almost beak-like, hence the name "snout fly." This long "snout" functions as a sheath for the fly's long mouthparts. With mouthparts extended, this fly can obtain nectar from tubular-shaped flowers such as those in the mint family. The abdomen is nearly as wide as it is long (wider than thorax). The color is black with pairs of yellow rectangles on segments 2 and 3. Larvae are thought to live and feed in the dung of large herbivores such as moose, caribou, bison, and beef cattle. However, the presence of the flies in locations without a source of dung indicates that the larvae can make do with other wet habitats that have high organic content.

Field Notes: When startled, these flies often will fly to a safe perch on the underside of a leaf, making them very difficult to keep track of and not easy to photograph.

Range: Eastern with some western records

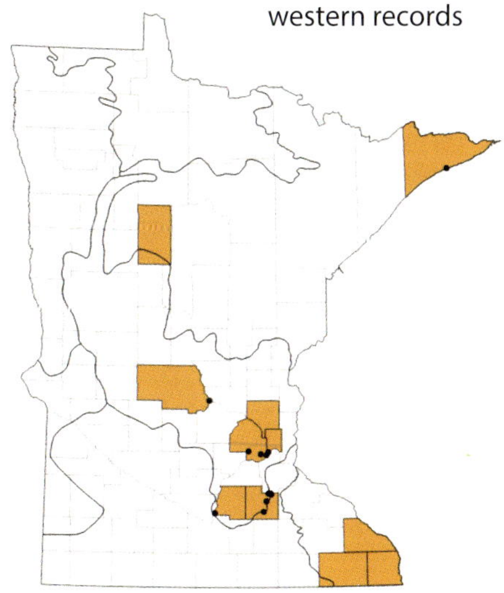

Status: Common

Flight Dates: May - August

Similar species: None.

Rhingia nasica

Flowers: The specialized mouthparts allow the adults to feed on a variety of flowers with deep nectaries, like the tubular flowers of mints.

Tribe Volucellini

Tribe Volucellini includes two genera in Minnesota.

Genus *Copestylum* :: Bromeliad Flies

These flies are closely associated with cactus. Two species are known from Minnesota, both from the far southwestern corner of the state.

Hairy-horned Bromeliad Fly

The Hairy-horned Bromeliad Fly (***Copestylum caudatum***) is a small, Nearctic flower fly with a western distribution. The silvery, appressed hairs on the fourth abdominal segment distinguish this species from other Copestylum species. Its arista is oddly thickened instead of plumose. It has a long, yellow face with a black median stripe. Two yellow spots at the rear of the thorax just in front of the scutellum. The larvae feed upon the decaying platyclades of cacti of the genus *Opuntia*.

Field Notes: There is a single specimen of this fly for Minnesota collected by Horace Telford in Luverne in July, 1933. There are several additional specimens collected in the South Dakota Badlands to the west.

Similar species: *Copestylum caudatum* is one of three nearly indistiguishable western species, the other two are *C. lentum* and *C. marginatum*.

Flowers: No Minnesota floral records. In Colorado, adult flies are known to visit aster and goldenrod.

Flight Dates: July

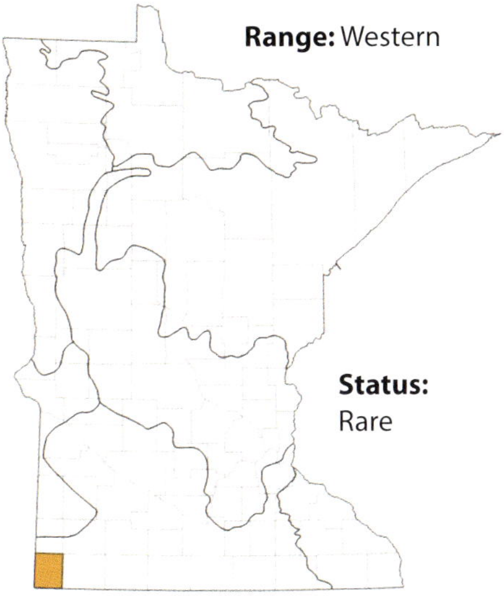

Range: Western

Status: Rare

Striped Bromeliad Fly

The Striped Bromeliad Fly (***Copestylum vittatum***) is a small, Nearctic flower fly. It has a long, yellow face with a black median stripe. Two yellow spots at the rear of the thorax just in front of the scutellum. Like *Volucella*, it has a plumose arista. Its bold yellow patterning seems to mimic that of the carder bees (tribe Anthidiini). The larvae feed upon the decaying platyclades of cacti of the genus *Opuntia*.

Field Notes: After the few records for Minnesota from the Blue Mounds area in 1922 and 1953 a new population was discovered at a rock outcropping with lots of cactus near Granite Falls in July, 2020.

Similar species: Most similar to *C. caudatum* but shouldn't be confused with other species.

Flowers: *Scrophularia* in Minnesota. *Helianthus, Prunus, Solidago*, and *Symphyotrichum* elsewhere.

Flight Dates: July - September

Range: Southern

Status: Rare

Genus *Volucella* :: Swiftwings

The genus *Volucella* includes large, fuzzy, bumblebee-like, species. The arista is plumose. The wing venation is distinct: the M1 vein curves backward where it meets the R4+5 vein. The face is triangular and projected downward. The larvae develop in the nests of social wasps and bumblebees where they are detritivores and predators of wasp larvae. **Note:** the *Volucella bombylans* species complex was only recently unraveled into three Nearctic species, two of which occur in Minnesota.

Yellow-faced Swiftwing

The Yellow-faced Swiftwing (***Volucella facialis***) is a large, Nearctic species. The arista is plumose. The wing venation is distinct: the M1 vein curves backward where it meets the R4+5 vein. The face is triangular and projected downward. The ground color of the face is yellow and has yellow pile covering it. The thorax is covered in long, yellow pile. The larvae develop in the nests of social wasps and bumble bees.

Field Notes: The abdomen pile is variable, often with orange pile on segment three but sometimes black. A mixed-boreal species often found on flowers associated with bog habitat.

Range: Northern

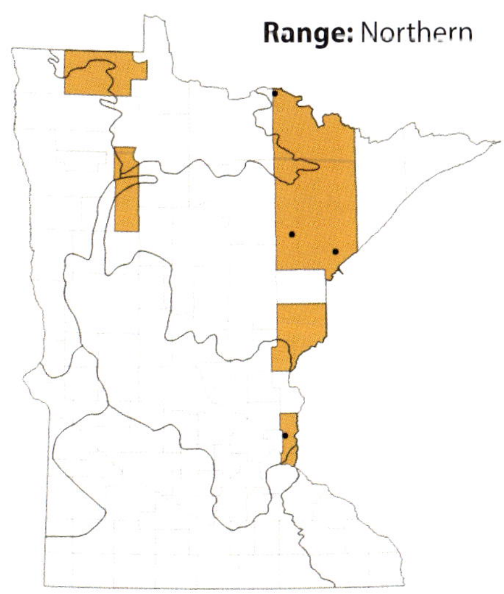

Status: Uncommon

Similar species: Very similar to Eastern Swiftwing. And both of these species are closely related to the European species *Volucella bombylans*. It could also be mistaken for some of the other bumblebee mimics. And is easily confused with actual bumblebees.

Flowers: No floral records for Minnesota. Elsewhere, it has been observed visiting brambles, bog laurel, and valerian.

Flight Dates: June - July

Eastern Swiftwing

The Eastern Swiftwing (***Volucella evecta***) is a large, Nearctic species. The arista is plumose. The wing venation is distinct: the M1 vein curves backward where it meets the R4+5 vein. The face is triangular and projected downward. The ground color of the face is black (often with sparse yellow pile). The thorax is covered in long, yellow pile. The larvae develop in the nests of social wasps and bumble bees.

Field Notes: J. Henri Fabre, in The Life of the Fly, pondered how the larvae of these flies could possibly survive inside a wasp nest: "What has it to make itself thus respected? Strength? Certainly not. It is a harmless creature, which the Wasp could rip open with a blow of her shears, while a touch of the sting would mean lightning death."

Range: Eastern

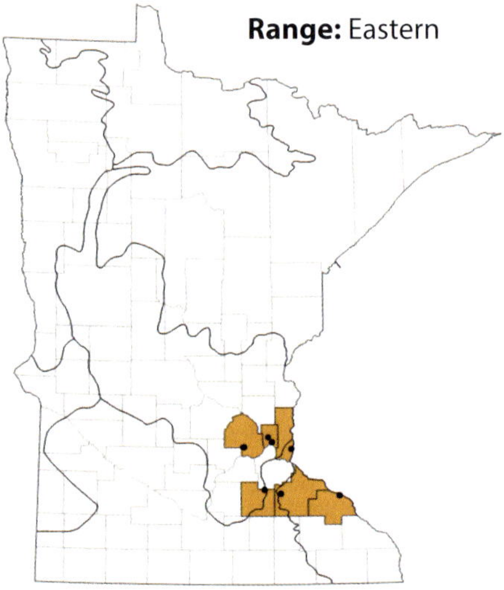

Similar species: Very similar to Yellow-faced Swiftwing. And both of these species are closely related to the European species *Volucella bombylans*. It could also be mistaken for some of the other bumblebee mimics. And is easily confused with actual bumblebees.

Flowers: Floral records for Minnesota include sumac, yarrow, and Indian hemp.

Flight Dates: June - July

Status: Uncommon

Chapter Five
MICRODONTINAE

MICRODONTINAE

Genera Summary: *Microdon, Mixogaster, Omegasyrphus*

Genus *Microdon* :: Ant Flies

The flies of genus *Microdon* are distinctly unusual flower flies, both morphologically and behaviorally. The R4+5 vein has an appendix that projects backward into cell r4+5. The larvae reside in ant nests where they prey upon ant eggs and larvae. Four species are known from Minnesota, with an additional three species possible.

Greater Ant Fly

The Greater Ant Fly (***Microdon manitobensis***) is a large, Nearctic flower fly. This fly has long, golden pile on the thorax, scutellum. The scutellum has two long points (calcars) that are difficult to see because of the thick pile. The abdomen has patches of white pile on each segment, giving it a patterned look, especially along the margin. The legs are black with white pile. The face is flat with white pile. The antennae are long. The larvae have been found in nests of *Formica* ants.

Field Notes: Ant flies perch on vegetation near to the ground in close proximity to ant colonies.

Similar species: Most similar to *Microdon megalogaster.*

Flowers: Atypical for flower flies, the adults apparently do not require pollen or nectar and, therefore, are not likely to be observed at flowers.

Flight Dates: June

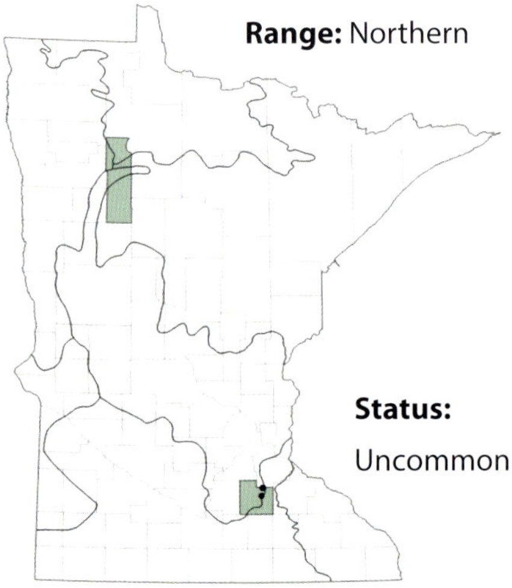

Range: Northern

Status: Uncommon

Black-bodied Ant Fly

The Black-bodied Ant Fly (***Microdon megalogaster***) is a large, Nearctic flower fly. This fly has long yellow pile with a slight metallic sheen on the thorax, scutellum, and the 1st and 2nd segments of the abdomen. The remainder of the abdomen, segments 3 through 5, is black, with short black pile. The scutellum has two, blunt points (calcars). The legs are black. The face is flat with white pile. The antennae are long. The larvae have been found in nests of Formica ants.

Field Notes: Ant flies perch on vegetation near to the ground, usually in close proximity to ant colonies.

Similar species: Most similar to *Microdon manitobensis.*

Flowers: Atypical for flower flies, the adults apparently do not require pollen or nectar and, therefore, are not likely to be observed at flowers.

Flight Dates: June

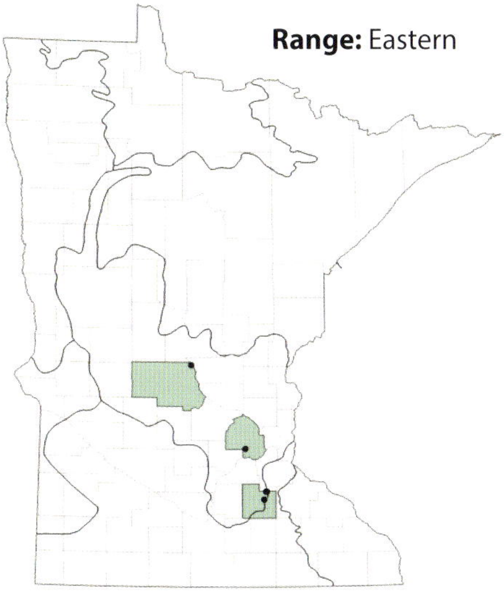

Range: Eastern

Status: Uncommon

Long-horned Ant Fly

The Long-horned Ant Fly (***Microdon tristis***) is a medium-sized, black-bodied Nearctic flower fly. This fly is almost entirely black with small amounts of short, golden pile. The thorax often has a purplish sheen and several dark stripes are visible on the dorsum. There is also, a large bare spot on thorax adjacent to the scutellum. The scutellum has two prominent points (calcars), directed back and slightly outward. The femurs are black and the tibia are covered in white pile. The antennae are long, with the flagellum about equal in length to the scape. The larvae have been found in *Camponotus* and *Formica* ant nests.

Field Notes: Slight variations in leg color, antennae length, and scutellum shape, led Horace Telford to describe the Minnesota variety as a new species, *Microdon robusta*, treated here as a synonym.

Similar species: Most similar to *Microdon globosus* and other dark *Microdon* species but could also be mistaken for a bot fly.

Flowers: Atypical for flower flies, the adults apparently do not require pollen or nectar and, therefore, are not likely to be observed at flowers.

Flight Dates: June

Range: Eastern (with a disjunct Western population)

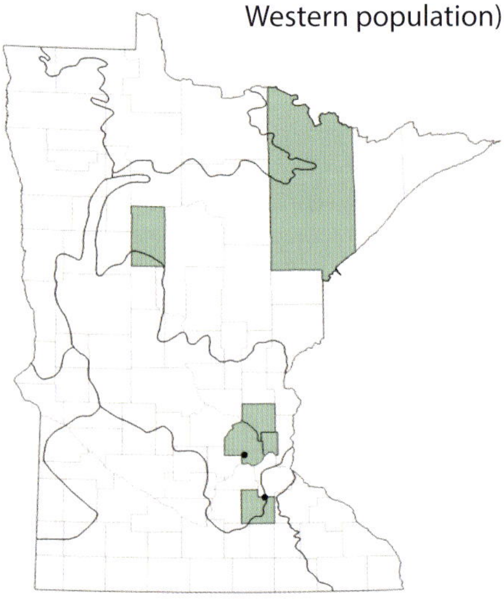

Status: Common

Globular Ant Fly

The Globular Ant Fly (***Microdon globosus***) is a medium-sized, brownish-orange Nearctic flower fly. Usually orange-brown in color, it is darker, almost black in Minnesota. The wing veins are darkened. The scutellum has two, triangular points (calcars), directed straight backward. The abdomen is very wide. The femurs are black and the tibia are covered in white pile. The antennae are long, with the flagellum longer than the scape and the arista greater than 1/2 the length of the flagellum. The basal tarsomeres on the hind legs are greatly enlarged. The larvae have been found in *Tapinoma sessile* ant nests.

Field Notes: Flies later in the year than other *Microdon* species. They tend to perch with their abdomen curled under, giving them a hunched-up look.

Similar species: Most similar to other *Microdon tristis* and other dark *Microdon* species.

Flowers: Adults are not known to visit flowers.

Flight Dates: June - September

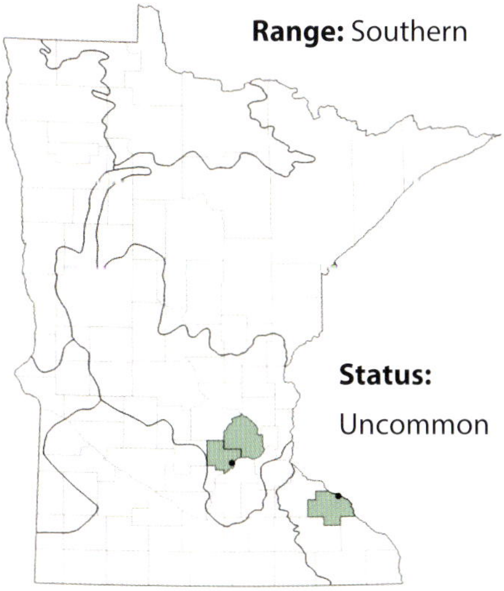

Range: Southern

Status: Uncommon

Additional Minnesota Species Not Included in the Guide

Microdon abditus: Eastern; similar to *M. globosus* but shorter flagellum
Microdon albicomatus: * Northern; short flagellum and very small calcars
Microdon cothurnatus: * Northern; medium flagellum and very small calcars

Genus *Mixogaster* :: Ant Flies

The flies of genus *Mixogaster* are small, curious flower flies. The wing venation and antennae shape are distinctive. The wing vein M1 has a short appendix that projects forward into cell r4+5. The antennae have long, rod-shaped flagellum. Their coloration and shape make them convincing wasp mimics. So far, only one of the four Nearctic species has been recorded in Minnesota.

Slender Ant Fly

The Slender Ant Fly (***Mixogaster breviventris***) is a medium-sized, Nearctic flower fly. The antennae have thin, elongate scapes and long, rod-shaped flagellum. The vertex is partly black and there is a black, horizontal stripe crossing the face just above the antennae. There are yellow, lateral stripes on the thorax. The biology of the larvae remains a mystery, possibly associated with odorous ants (subfamily Dolichoderinae).

Field Notes: John Haarstad noted that this species was quite abundant where he collected it at Cedar Creek Ecosystem Science Reserve. Curiously, this remains the only known location in Minnesota.

Similar species: Most similar to other *Mixogaster* but could, perhaps, be mistaken for *Xanthogramma*.

Flowers: Adults are not known to visit flowers.

Flight Dates: June

Range: Eastern

Status: Rare

Genus *Omegasyrphus* :: Ant Flies

The flies of genus *Omegasyrphus* have narrow abdomens and distinct wing venation. The second abdominal segment is noticeably more convex and rounded than other segments. The R4+5 vein has an appendix that projects backward into cell r4+5. One species is known from Minnesota.

Eastern Spot-winged Ant Fly

The Eastern Spot-winged Ant Fly (***Omegasyrphus coarctatus***) is a small, Nearctic flower fly. The abdomen is narrow and greenish-bronze in color. The R4+5 vein has an appendix that projects into cell r4+5. This appendix and the other crossveins are darkened, giving the wing a spotted appearance. The larvae are associated with ants of genera *Aphaenogaster* and *Monomorium*.

Field Notes: Known from specimens collected at two locations in western Minnesota. One location was Coleman WMA southeast of Morris in 1935. It would be wonderful to learn that that population persists.

Similar species: Shouldn't be confused with other flower flies but might be mistaken for a soldier fly (family Stratiomyidae).

Flowers: Adults are not known to visit flowers.

Flight Dates: June

Range: Southern

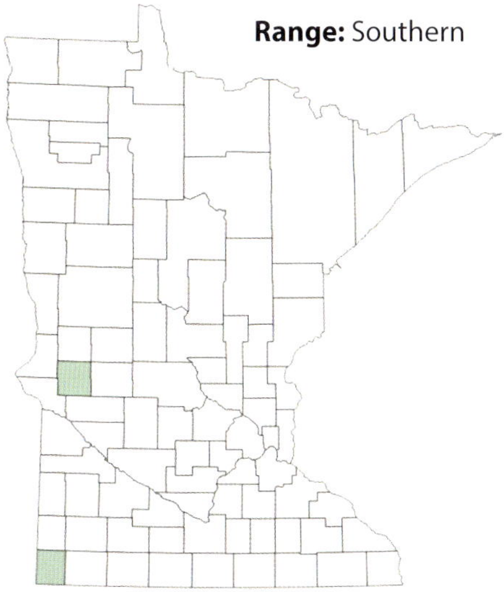

Status: Rare

GLOSSARY: ANATOMICAL TERMS

Abdomen
The body of the insect attached to the rear of the thorax.

Alula
A broad lobe at the proximal posterior margin of the wing stalk of Diptera. Also termed the axillary lobe.

Anepimeron
A central plate of the mesothorax, located below and forward of the wing.

Anepisternum
A central plate of the mesothorax, located directly below wing and behind anepimeron.

Antennae
A pair of appendages attached to the face between the eyes. They are sense organs used to detect airborne scents and air movements. A flower fly antenna has three segments: the flagellum, the scape, and the pedicel.

Antennomere
A segment of an antenna. The term antennomere is used in particular when the segments are fairly uniform, as in filiform antennae, but it also may be used in referring to segments of odd sizes, shapes and functions, such as the scape and pedicel.

Anterior
In front of. Opposite posterior.

Arista
Long, stiff, bristle-like hair attached to the base of the flagellum, the third segment of the antenna.

Calcars
An elongated extension of the exoskeleton, as often found on the scutellum of Microdon.

Calypters
Small membranous flaps or lobes that are located at the base of the wing. Also known as squamae.

Cell
The central area enclosed by wing veins. It can be closed by veins or open.

Compound Eye
An eye consisting of a large number of individual photoreceptor units or ommatidia.

Costa or Costal area
The leading edge of the wing.

Coxa
First leg segment, between body and trochanter.

Dichoptic
Compound eyes separated by a gap at the top of the head.

Dorsal
The upper surface of the body or body part.

Face
Area between the eyes, below the base of antennae.

Fascia (plural Fasciae)
A color pattern with a broad band.

Femur
Third leg segment, usually the longest and largest part of the leg, located between trochanter and tibia.

Flagellum
The end of the antenna, the third segment connected to the pedicel.

Frons
Frontal area of an insect's head, between the eyes and above the antennae.

Gena
The area of the face below the compound eyes, the insect equivalent to human cheeks.

Glabrous
Smooth, without hairs or scales.

Haltere
Vestigial hind wing that acts as a sensory organ of balance; a unique trait of all Diptera.

Holoptic
Eyes that meet at the top of the head.

Humerus (plural Humeri)
A plate on the front corner of the dorsal part of the mesothorax, also known as postpronotum.

Hyaline
Transparent, like glass.

Katepisternum
A central plate of the mesothorax, located below the anepimeron and anepisternum and above the middle coxa.

Mesolegs
The middle pair of legs.

Mesothorax
The middle part of the thorax to which is attached the wings and the mesolegs.

Metalegs
The hind pair of legs, the rear legs.

Metathorax
The hindmost part of the thorax to which is attached the abdomen, the haltere, the scutellum, and the metalegs.

Metasternum
The ventral area in front of the rear legs.

Ocellus (plural Ocelli)
Simple eye; flower flies have three arranged in a triangle between the compound eyes on the top of the head.

Ommatidia
Photoreceptor with a tiny, single lens, the individual component of compound eyes.

Ovipositor
Tubular structure by which the female fly lays her eggs.

Pedicel
The second segment of the antenna, between scape and flagellum.

Plumose
Covered in hairs like a feather.

Postalar Callus

A portion of the scutum separated by a groove and slightly raised, located directly behind the wings.

Postpronotum

See Humerus.

Proboscis

Mouthparts that take in liquids and solids.

Prolegs

The foremost pair of legs, the front legs.

rothorax

The anterior part of the thorax to which is attached the head and the prolegs.

Puparium

The exoskeleton of the final instar inside of which the pupa develops.

Scape

The first segment of the antenna, attached to the face in the antennae pit.

Sclerites

Definable section of the exoskeleton.

Scutellum

The distinct, half-moon portion of the mesonotum, located at the rear of the thorax, above the first segment of the abdomen.

Spiracle

Respiratory openings on the thorax and abdomen that allow air to enter the trachea.

Sternite

Chitinous plates on the underside of each abdominal segment.

Tarsus (plural Tarsi)

The last leg segment which consists of five small segments, the fifth and final segment has a pair of claws.

Tergite

The chitinous plate forming the dorsal surface of each abdominal segment.

Thorax

The part of the body between the head and the abdomen. It has three parts: the prothorax, the mesothorax, and the metathorax.

Tibia

Fourth leg segment, usually long and thin, located between femur and tarsus.

Trochanter

Second leg segment, between coxa and femur.

Vein

Hollow structures formed from the coupling of the upper and lower walls of the wing. They provide both rigidity and flexibility to the wing.

Vena Spuria

The spurious vein, a thickened fold in the center of the wing that looks like a vein; a unique characteristic of flower flies.

Vertex

Apex of the head, between the compound eyes, including the ocelli.

GLOSSARY: ECOLOGICAL TERMS

Adventive
Introduced species not fully naturalized.

Aphidophagy
Feeding on aphids.

Batesian Mimicry
Type of mimicry where an unrelated species acquires the warning colors and markings of a species well-defended from predators.

Bivoltine
Two full life cycles per year.

Boreal
Biome characterized by coniferous forests.

Diapause
A period of physiologically enforced dormancy between periods of activity.

Dendrolimnetobiont
Larvae develop exclusively in rot holes in living trees.

Dendrolimnetophil
Larvae will sometimes develop in rot holes but also may use other aquatic habitats.

Entomophagy
Feeding on other insects.

Holarctic
Geographical region that includes the Nearctic and Palaearctic.

Instars
A stage in the life of an insect between two successive molts.

Mixed Boreal
Biome characterized by mixed deciduous-conifer forests.

Müllerian Mimicry
Müllerian mimicry occurs when two or more well-defended species mimic each other's warning signals.

Multivoltine
Multiple life cycles per year.

Nearctic
Geographical region that includes North America.

Oviposition
The act of laying eggs.

Palaearctic — Geographical region that includes Europe and Asia north of the Himalayas.

Phytophagy
Feeding on plants.

Pollinophagy
Feeding on pollen.

Saprophagy
Feeding on decaying organic matter.

Semivoltine
Species that require two or more years to complete their life cycles.

Status
Measure of species population stability.

Univoltine
Species that complete their life cycle once in a single year.

Xylophagy — Feeding on wood.

REFERENCES

Ball, Stuart and Roger Morris. 2015. *Britain's Hoverflies: A Field Guide. WildGuides*. Princeton University Press.

Bankowska, Regina. 1998. Syrphid flies (Syrphidae, Diptera) from North Korea collected by Polish expeditions. Part I (Milesiinae, Microdontinae). *Fragmenta Faunistica, Warszawa. 41*: 167-182.

Виолович Н. А. «Сирфиды Сибири» (Diptera, Syrphidae). Определитель. — Новосибирск, Наука, 1983

Deyrup, M. 2017. *Ants of Florida: Identification and Natural History*. CRC Press. Boca Raton, Florida.

Dirzo R., Young H. S., Galetti M., Ceballos G., Isaac N.J., Collen B. 2014. Defaunation in the Anthropocene. *Science 345*:401–6

Fluke, C. L. and F. M. Hull. 1946. Syrphid Flies of the Genus Cheilosia, subgenus Chilomyia in North America (Part II). *Transactions of the Wisconsin Academy of Sciences, Arts, and Letters 36*: 327-347.

Fluke, C. L. and F. M. Hull. 1947. The Cartosyrphus Flies of North America (Syrphidae). *Transactions of the Wisconsin Academy of Sciences, Arts, and Letters 37*: 221-263.

Gilbert F. 2005. The evolution of imperfect mimicry in hoverflies. Insect Evolutionary Ecology: *Proc. Royal Ent. Society's 22nd Symposium*: 231–288.

Grissell, Eric. 2010. *Bees, Wasps, and Ants: The Indispensable Role of Hymenoptera in Gardens*. Timber Press, Portland.

Hull, F. M. and Fluke Jr., C. L. 1950. The Genus Cheilosia Meigen (Diptera, Syrphidae): The subgenera Cheilosia and Hiatomyia. *Bulletin of the American Museum of Natural History Vol.94*: Article 6, 301-401. New York.

iNaturalist Phenology Chart. Available from https://www.inaturalist.org. Accessed [12/23/2020].

Locke, M. M. and Skevington, J. H.2013. Revision of Nearctic Dasysyrphus Enderlein (Diptera: Syrphidae). *Zootaxa, [S.l.], v. 3660*, n. 1, p. 1–80, May.

Mabey, Richard. 1980. *The Common Ground: A Place for Nature in Britain's Future*. Hutchinson.

Marren, Peter. 2002. *Nature Conservation: A Review of the Conservation of Wildlife in Britain 1950 - 2001*. New Naturalist Series. Harper Collins.

McAlister, Erica. 2017. T*he Secret Life of Flies*. Natural History Museum, London.

Mengual, X., G. Stahls, and S. Rojo. 2015. "Phylogenetic relationships and taxonomic ranking of pipizine flower flies (Diptera: Syrphidae) with implications for the evolution of aphidophagy." *Cladistics* 1-18.

Mutin, V. A. and Barkalov, A. V. 1999. "Family Syrphidae-Hover Flies. A Key to the Insects of the Russian Far East." Vol. 6: No. 1: pp. 342–500. Dalnauka, Vladivostok. [in Russian]

Penney H.D., Hassall C., Skevington J.H., Abbott K.R., Sherratt T.N. 2012. A comparative analysis of the evolution of imperfect mimicry. *Nature 483*: 461–464.

Reemer, M. 2013. Review and phylogenetic evaluation of associations between Microdontinae (Diptera: Syrphidae) and ants (Hymenoptera: Formicidae). *Psyche: A Journal of Entomology*: ID:538316.

Reemer M., Ståhls G. 2013. Generic revision and species classification of the Microdontinae (Diptera, Syrphidae). *ZooKeys 288*: 1–213.

Redfern, Margaret. 2011. *Plant Galls*. The New Naturalist Library. Collins, London.

Rotheray G.E. 1993. Colour guide to hoverfly larvae (Diptera, Syrphidae) in Britain and Europe. *Dipterists Digest No. 9*. 155 pp.

Rotheray, Graham E. and Francis Gilbert. 2011. *The Natural History of Hoverflies*. Forrest Text.

Shah, G. M., U. Jan and A. A. Wachkoo. 2014. A Checklist of Hoverflies (Diptera: Syrphidae) in the Western Himalaya, India. *Acta Zoologica Academiae Scientiarum Hungaricae 60*(4).

REFERENCES

Skevington Jeffrey H. and D. K. Yeates. 2000. Phylogeny of the Syrphoidea (Diptera) inferred from mtDNA sequences and morphology… *Molecular Phylogenetics and Evolution 16*: 212–224.

Skevington, Jeffrey H., Michelle M. Locke, Andrew D. Young, Kevin Moran, William J. Crins, and Stephen A. Marshall. 2019. *Field Guide to the Flower Flies of Northeastern North America*. Princeton University Press.

Ståhls G., Hippa H., Rotheray G., Muona J., Gilbert F. 2003. Phylogeny of Syrphidae (Diptera) inferred from combined analysis of molecular and morphological characters. *Systematic Entomology 28*: 433–450.

Telford, Horace S. 1939. *The Syrphidae of Minnesota*. University of Minnesota Agricultural Experiment Station.

Telford, Horace S. 1975. Records of Flower Flies (Syrphidae: Diptera) of Washington State. *Melanderia Vol. 22*.

Young, Andrew D.; Marshall, Stephen A.; Skevington, Jeffrey H.. Revision of Platycheirus Lepeletier and Serville (Diptera: Syrphidae) in the Nearctic north of Mexico. *Zootaxa, [S.l.], v. 4082*, n. 1, p. 1–317, feb. 2016

Veen, Mark P. 2010. *Hoverflies of Northwest Europe: Identification Keys to the Syrphidae*. KNNV Publishing.

Vockeroth J.R. 1992. The flower flies of the subfamily Syrphinae of Canada, Alaska, and Greenland, Diptera: Syrphidae. *The insects and arachnids of Canada*, Pt. 18. Ottawa: Agriculture Canada. 456 pp.

White, Denis. 2020. *Ecological Regions of Minnesota: Level III and IV maps and descriptions* [map]. United States Environmental Protection Agency. https://www.epa.gov/eco-research/ecoregion-download-files-state-region-5#pane-21

WEBSITES AND ONLINE RESOURCES

Canadian National Collection of Insects, Arachnids and Nematodes

Nearctic Syrphidae Checklist
http://www.canacoll.org/Diptera/Staff/Skevington/Syrphidae/Syrphidae_Nearctic_Checklist.htm

Links to Species Pages for the Field Guide to the Syrphidae of Ontario
http://www.canacoll.org/Diptera/Staff/Skevington/Syrphidae/speciespages.htm

University of British Columbia Spencer Entomological Collection: Syrphidae
https://www.zoology.ubc.ca/~biodiv/entomology/main/Diptera/Syrphidae/

Key to the Insects of Russian Far East: Vol. VI Diptera and Syphonaptera Pt. 1 (Vladivostok, 1999)
http://www.online-keys.net/sciaroidea/add01/opred_nasek_daln_vostka1999_6_1.pdf

University of Minnesota Cedar Creek Ecosystem Science Reserve Insect Photo Album: Diptera, Syrphidae
https://www.cedarcreek.umn.edu/insects/orderpages/029pg-syrphidae.html

Diptera.info
https://diptera.info/news.php

BugGuide
https://bugguide.net/

iNaturalist
https://www.inaturalist.org

Field/Photo ID for Flies
https://sites.google.com/view/flyguide

Japanese Syrphidae
http://syrphidae.a.la9.jp

PHOTOGRAPHY CREDITS

All photographs were taken by the author, ©Scott King, with the exception of the following:

E. Christie Butler: Teuchocnemis lituratus, p. 186.

Jason Michael Crockwell: Common thintail (two photos), p. 88.

Rob Curtis: Black-bristled logsitter, p. 121; Blera analis, p. 169; Copestylum vittatum, p. 212.

Denis Doucet: Confusing conifer fly (two photos) p. 74; black-horned smoothtail (two photos), p. 77; black-legged gossamer, p. 84; Spilomyia fusca, p. 177; Chalcosyrphus curvaria, p. 193.

PHOTOGRAPHY CREDITS

David Enrique: Hair-eyed Halfband, p. 86; Brachyopa caesariata, p. 120.

Kimberly R Fleming: Williston's wasp fly larva, p. 133.

Solomon Hendrix: American dainty, p. 59.

Heather Holm: Syritta pipiens, p. 2; drone fly on goldenrod, p. 3; chapter one cover photo, p. 12; Eristalis tenax, p. 18; dichoptic eyes, p. 20; compound eyes, p. 20; Rhingia nasica, pp. 22 and 210; Dolichovespula arenaria, p. 28; Bombus rufocinctus, p. 29; Spilomyia sayi, p. 29; Ancistrocerus catskill, p. 29; Eristalis dimidiata, p. 30; larva, p. 31; pollen consumption, p. 33; Helophilus fasciatus, p. 34; larva, p. 36; jagged ambush bug, p. 39; Ectemnius maculosus, p. 39; Bembix americana, p. 39; reconstructed prairie, p. 41; flower fly on rose, p. 42; raingarden, p. 43; Claytonia virginica, p. 46; mucksucker, p. 47; Microdon, p. 49; garden, p. 50; chapter two cover photo, p. 55; Sphaerophoria contigua, p. 58; Paragus haemorrhous, p. 69; Epistrophella larva, p. 78; cobalt hover fly (two photos), p. 91; sumac gall pithead (two photos), p. 110; Eristalinae cover, p. 113; Orthonevra nitida (two photos), p. 124; Orthonevra pictipennis, p. 126; Mylolepta strigilata (two photos), p. 127; Neoasicia metallica (two photos), p. 128; Sphegina campanulata, p. 130; Ceriana abbreviata on flower, p. 132; Eristalis, p. 135; Eristalis anthophorina, p. 137; Eristalis flavipes, p. 142; Helophilus fasciatus, p. 152; Mallota bautias on rose, p. 160; Eumeris funeralis on flower, p. 167; Somula decora, p. 168; Spilomyia sayi on goldenrod, p. 175; Spilomyia alcimus (four photos), p. 176; Spilomyia longicornis on flower, p. 178; Temnostoma alternans mating, p. 181; Temnostoma balyras on flower, p. 183; Syritta pipiens on flower, p. 187; Chalcosyrphus chalybeus on flower, p. 192; Volucella evecta on flower, p. 214; Microdontinae chapter cover, p. 215; Microdon tristis (two photos), p. 218.

Angella Moorehouse: Xylota angustiventris, p. 197.

Tom Murray: Eastern smoothleg (two photos), p. 106; common copperback, p. 208.

Janet Nelson: Orange-spotted drone fly, p. 137; Spilomyia fusca (three photos), p. 177.

Bob Noble: Large-spotted halfband, p. 85.

Jim Parkin: Shutterstock photo ID: 56899537, Backyards, p. 40.

John Rosenfeld: Pearly sedgesitter, p. 65.

Stephen Mlodinow: Copestylum caudatum, p. 211.

Stephen Rudolph: Shutterstock photo ID: 53567515, Arborist, p. 44.

Harvey Schmidt: Xylota flavitibia (two photos), p. 196.

Jeff Skevington: Arctosyrphus willingii, p.148.

Stuart Tingley: Canadian potterfly, p. 76; common loopwing aphideater, p. 82; large-spotted halfband, p. 85; four-spotted pithead, p. 108.

Released into the public domain: Black-margined flower fly, p. 98; Milesia virginiensis, p. 174.

CHECKLIST OF MINNESOTA FLOWER FLIES

A checklist of flower fly species recorded from Minnesota is reported based upon literature records, entomological collections, and citizen scientist observations. Currently 72 genera and 187 species are known from the state [with at least an additional 4 genera and 50 species expected or at least probable—these hypothetical species are marked with an asterisk]. This checklist provides a long-needed update to the work of Horace Telford who published a synopsis of the flower flies in 1932.

The synonyms appended after the checklist include those names used in Telford's original publication on Minnesota flower flies, Cedar Creek and University of Minnesota specimens, and those few names that differ from online resources at iNaturalist and BugGuide. Skevington et.al. (2019) contains a more complete listing of synonyms.

Hypothetical species are marked with an asterisk. Several names have not been tracked down to current usage and these are marked with a question mark.

SUBFAMILY SYRPHINAE

- ☐ *Baccha cognata* (Loew 1863)
- ☐ *Melanostoma mellinum* (Linnaeus 1758)
- ☐ *Platycheirus angustatus* (Zetterstedt 1843)
- ☐ *Platycheirus clypeatus* (Meigen 1822) *
- ☐ *Platycheirus confusus* (Curran 1925)
- ☐ *Platycheirus hyperboreus* (Staeger 1845)
- ☐ *Platycheirus immarginatus* (Zetterstedt 1849)
- ☐ *Platycheirus luteipennis* (Curran 1925) *
- ☐ *Platycheirus modestus* Ide 1926
- ☐ *Platycheirus nearcticus* Vockeroth 1990
- ☐ *Platycheirus nodosus* Curran 1923 *
- ☐ *Platycheirus obscurus* (Say 1824)
- ☐ *Platycheirus pictipes* (Bigot 1884)
- ☐ *Platycheirus quadratus* (Say 1823)
- ☐ *Platycheirus scambus* (Stæger 1843)
- ☐ *Platycheirus scutatus* (Meigen 1822) *
- ☐ *Platycheirus thompsoni* Vockeroth 1990
- ☐ *Platycheirus varipes* Curran 1923 *
- ☐ *Platycheirus granditarsis* (Forster 1771)
- ☐ *Platycheirus rosarum* (Fabricius 1787)
- ☐ *Paragus haemorrhous* Meigen 1822
- ☐ *Paragus angustifrons* Loew 1863
- ☐ *Paragus angustistylus* Vockeroth 1986 *
- ☐ *Paragus bispinosus* Vockeroth 1986 *
- ☐ *Paragus variabilis* Vockeroth 1986 *
- ☐ *Chrysotoxum derivatum* Walker 1849
- ☐ *Chrysotoxum flavifrons* Macquart 1842
- ☐ *Chrysotoxum plumeum* Loew 1864
- ☐ *Chrysotoxum pubescens* Loew 1864
- ☐ *Dasysyrphus intrudens* (Osten Sacken 1877)
- ☐ *Dasysyrphus laticaudus* (Curran, 1925)*
- ☐ *Dasysyrphus venustus* (Meigen 1822)
- ☐ *Didea fuscipes* Loew 1863
- ☐ *Doros aequalis* Loew, 1863
- ☐ *Epistrophe grossulariae* (Meigen 1822)
- ☐ *Epistrophe nitidicollis* (Meigen 1822) *
- ☐ *Epistrophe xanthostoma* (Williston 1887)
- ☐ *Epistrophella emarginata* (Say 1823)
- ☐ *Eupeodes volucris* Osten Sacken 1877
- ☐ *Eupeodes americanus* (Wiedemann 1830)

- ☐ *Eupeodes latifasciatus* (Macquart 1829)
- ☐ *Eupeodes perplexus* (Osburn 1910)
- ☐ *Eupeodes flukei* (Jones 1917) *
- ☐ *Eupeodes luniger* (Meigen 1822) *
- ☐ *Eupeodes pomus* (Curran 1921)
- ☐ *Hypocritanus fascipennis* (Wiedemann 1830)
- ☐ *Lapposyrphus lapponicus* (Zetterstedt 1838)
- ☐ *Leucozona americana* Curran 1923 *
- ☐ *Megasyrphus laxus* (Osten Sacken 1875)
- ☐ *Melangyna fisherii* (Walton 1911)
- ☐ *Melangyna lasiophthalma* (Zetterstedt 1843)
- ☐ *Melangyna umbellatarum* (Fabricius 1794) *
- ☐ *Meligramma triangulifera* (Zetterstedt 1843)
- ☐ *Meliscaeva cinctella* (Zetterstedt 1843)
- ☐ *Parasyrphus genualis* (Williston 1887)
- ☐ *Parasyrphus nigritarsis* (Zetterstedt 1843) *
- ☐ *Parasyrphus vockerothi* Thompson 2012
- ☐ *Pelecinobaccha costata* (Say, 1829)
- ☐ *Scaeva affinis* Say, 1823
- ☐ *Sphaerophoria abbreviata* Zetterstedt 1849 *
- ☐ *Sphaerophoria asymmetrica* Knutson 1972
- ☐ *Sphaerophoria brevipilosa* Knutson 1972 *
- ☐ *Sphaerophoria contigua* Macquart 1847
- ☐ *Sphaerophoria novaeangliae* Johnson 1916 *
- ☐ *Sphaerophoria philanthus* (Meigen 1822)
- ☐ *Sphaerophoria pyrrhina* Bigot 1884 *
- ☐ *Syrphus attenuatus* Hine, 1922 *
- ☐ *Syrphus knabi* Shannon 1916
- ☐ *Syrphus opinator* Osten Sacken, 1877
- ☐ *Syrphus rectus* Osten Sacken 1875
- ☐ *Syrphus ribesii* (Linnaeus 1758)
- ☐ *Syrphus sexmaculatus* (Zetterstedt, 1838)
- ☐ *Syrphus torvus* Osten Sacken 1875
- ☐ *Syrphus vitripennis* Meigen 1822
- ☐ *Xanthogramma flavipes* (Loew 1863)
- ☐ *Allograpta obliqua* (Say 1823)
- ☐ *Toxomerus geminatus* (Say 1823)
- ☐ *Toxomerus marginatus* (Say 1823)
- ☐ *Toxomerus politus* (Say 1823)

SUBFAMILY PIPIZINAE

- ☐ *Heringia canadensis* Curran 1921 *
- ☐ *Heringia salax* (Loew, 1866)
- ☐ *Neocnemodon calcarata* (Loew, 1866)
- ☐ *Neocnemodon carinata* (Curran, 1921)
- ☐ *Neocnemodon coxalis* (Curran 1921)
- ☐ *Neocnemodon elongata* (Curran 1921)
- ☐ *Neocnemodon intermedia* (Curran 1921)
- ☐ *Neocnemodon pisticoides* (Williston, 1887)
- ☐ *Neocnemodon rita* (Curran 1921) *
- ☐ *Pipiza atrata* Curran 1922
- ☐ *Pipiza femoralis* Loew 1866
- ☐ *Pipiza macrofemoralis* Curran 1921 *
- ☐ *Pipiza puella* Williston 1887
- ☐ *Pipiza quadrimaculata* (Panzer 1804) *
- ☐ *Trichopsomyia apisaon* Walker 1849
- ☐ *Trichopsomyia banksi* (Curran 1921) *
- ☐ *Trichopsomyia pubescens* (Loew, 1863) *
- ☐ *Trichopsomyia recedens* (Walker 1852)

- ☐ *Brachyopa caesariata* Miranda, 2020
- ☐ *Brachyopa notata* Osten Sacken 1875
- ☐ *Brachyopa vacua* Osten Sacken 1875
- ☐ *Chrysogaster antitheus* Walker 1849
- ☐ *Chrysosyrphus latus* (Loew 1863)
- ☐ *Hammerschmidtia rufa* (Fallén 1817) *
- ☐ *Myolepta nigra* (Loew 1872)
- ☐ *Myolepta strigilata* (Loew, 1872)
- ☐ *Orthonevra nitida* (Wiedemann 1830)
- ☐ *Orthonevra pictipennis* (Loew 1863)
- ☐ *Orthonevra pulchella* (Williston 1887)
- ☐ *Neoascia metallica* (Williston 1882)
- ☐ *Neoascia tenur* Harris 1780
- ☐ *Neoascia globosa* (Walker, 1849)
- ☐ *Neoascia* sp. (undescribed species 1 in Skevington et al, 2019)
- ☐ *Neoascia subchalybea* Curran 1925
- ☐ *Sphegina campanulata* Robertson 1901
- ☐ *Sphegina petiolata* Coquillett 1910
- ☐ *Sphegina rufiventris* Loew 1863
- ☐ *Sphegina brachygaster* Hull 1935
- ☐ *Sphegina flavimana* Malloch 1922
- ☐ *Sphegina lobata* Loew 1863
- ☐ *Callicera erratica* (Walker 1849) *
- ☐ *Ceriana abbreviata* Loew 1864
- ☐ *Sphiximorpha willistoni* (Kahl 1897)
- ☐ *Eristalinus aeneus* (Scopoli 1763)
- ☐ *Eristalis anthophorina* (Fallén 1817)
- ☐ *Eristalis arbustorum* (Linnaeus 1758)
- ☐ *Eristalis brousii* Williston 1882
- ☐ *Eristalis cryptarum* (Fabricius 1794)
- ☐ *Eristalis dimidiata* Wiedemann 1830
- ☐ *Eristalis flavipes* Walker 1849
- ☐ *Eristalis interrupta* (Poda 1761)
- ☐ *Eristalis obscura* Loew 1866
- ☐ *Eristalis stipator* Osten Sacken 1877
- ☐ *Eristalis transversa* Wiedemann 1830
- ☐ *Eristalis tenax* (Linnaeus 1758)
- ☐ *Helophilus fasciatus* Walker 1849
- ☐ *Helophilus hybridus* Loew 1846
- ☐ *Helophilus lapponicus* Wahlberg 1844 *
- ☐ *Helophilus latifrons* Loew 1863
- ☐ *Helophilus obscurus* Loew 1863
- ☐ *Arctosyrphus willingii* (Smith, 1912)
- ☐ *Anasimyia anausis* (Meigen, 1822)
- ☐ *Anasimyia bilinearis* (Williston 1887)
- ☐ *Anasimyia chrysostomus* (Wiedemann)
- ☐ *Eurimyia stipata* (Fabricius, 1787)
- ☐ *Polydontomyia curvipes* (Wiedemann, 1830)
- ☐ *Mallota bautias* (Walker 1849)
- ☐ *Mallota illinoensis* Robertson, 1901 *
- ☐ *Mallota posticata* Fabricius 1805
- ☐ *Parhelophilus brooksi* Curran 1927
- ☐ *Parhelophilus laetus* (Loew 1863)
- ☐ *Parhelophilus obsoletus* (Loew 1863)
- ☐ *Parhelophilus porcus* (Walker 1849)
- ☐ *Parhelophilus rex* Curran & Fluke 1922
- ☐ *Sericomyia bifasciata* Williston 1887 *
- ☐ *Sericomyia chrysotoxoides* Macquart 1842
- ☐ *Sericomyia lata* (Coquillett 1907)
- ☐ *Sericomyia militaris* Walker 1849
- ☐ *Sericomyia sexfasciata* Walker 1849 *
- ☐ *Sericomyia transversa* Osburn 1926
- ☐ *Sericomyia vockerothi* Skevington 2012
- ☐ *Eumerus funeralis* Meigen 1822
- ☐ *Eumerus strigatus* (Fallén 1817) *
- ☐ *Merodon equestris* (Fabricius, 1794)
- ☐ *Blera analis* (Macquart, 1842)
- ☐ *Blera armillata* (Osten Sacken 1875) *
- ☐ *Blera badia* (Walker 1849)
- ☐ *Blera confusa* Johnson, 1913
- ☐ *Blera nigra* (Williston 1887) *

- ☐ *Somula decora* Macquart 1847
- ☐ *Criorhina nigriventris* Walton 1911
- ☐ *Criorhina verbosa* (Walker 1849)
- ☐ *Sphecomyia vittata* (Wiedemann 1830)
- ☐ *Milesia virginiensis* (Drury 1773)
- ☐ *Spilomyia alcimus* (Walker 1849)
- ☐ *Spilomyia fusca* Loew 1864
- ☐ *Spilomyia longicornis* Loew 1872
- ☐ *Spilomyia sayi* (Goot 1964)
- ☐ *Pterallastes thoracicus* Loew 1863
- ☐ *Temnostoma alternans* Loew 1864
- ☐ *Temnostoma balyras* (Walker 1849)
- ☐ *Temnostoma barberi* Curran 1939
- ☐ *Temnostoma excentrica* (Harris 1862)
- ☐ *Temnostoma obscurum* Loew 1864
- ☐ *Temnostoma trifasciatum* Robertson 1901 *
- ☐ *Temnostoma venustum* Williston 1887 *
- ☐ *Teuchocnemis lituratus* (Loew 1863)
- ☐ *Syritta pipiens* (Linnaeus 1758)
- ☐ *Tropidia calcarata* Williston 1887 *
- ☐ *Tropidia mamillata* Loew 1861 *
- ☐ *Tropidia quadrata* (Say 1824)
- ☐ *Brachypalpus oarus* (Walker 1849)
- ☐ *Brachypalpus femoratus* (Williston 1882) *
- ☐ *Chalcosyrphus anthreus* (Walker 1849) *
- ☐ *Chalcosyrphus chalybeus* (Wiedemann 1830)
- ☐ *Chalcosyrphus curvaria* (Curran 1941)
- ☐ *Chalcosyrphus inarmatus* (Hunter 1897)
- ☐ *Chalcosyrphus libo* (Walker 1849)
- ☐ *Chalcosyrphus nemorum* (Fabricius 1805)
- ☐ *Chalcosyrphus nigromaculatus* (Jones 1917) *
- ☐ *Chalcosyrphus piger* (Fabricius 1794)
- ☐ *Chalcosyrphus vecors* (Osten Sacken 1875)
- ☐ *Xylota angustiventris* Loew 1866
- ☐ *Xylota annulifera* Bigot 1884
- ☐ *Xylota confusa* Shannon 1926
- ☐ *Xylota flavifrons* Walker 1849 *
- ☐ *Xylota flavitibia* Bigot 1884
- ☐ *Xylota hinei* (Curran 1941)
- ☐ *Xylota naknek* Shannon 1926
- ☐ *Xylota quadrimaculata* Loew 1866
- ☐ *Xylota subfasciata* Loew 1866
- ☐ *Xylota tuberculata* (Curran 1941)
- ☐ *Cheilosia comosa* (Loew 1863)
- ☐ *Cheilosia cottrelli* Telford 1939 *
- ☐ *Cheilosia cynoprosopa* Hull & Fluke 1950
- ☐ *Cheilosia hunteri* (Curran 1922)
- ☐ *Cheilosia lasiophthalmus* Williston 1882 *
- ☐ *Cheilosia latrans* (Walker 1849)
- ☐ *Cheilosia leucoparea* (Loew 1863) *
- ☐ *Cheilosia orilliaenis* (Curran, 1922)
- ☐ *Cheilosia rita* (Curran 1922)
- ☐ *Cheilosia shannoni* (Curran 1923) *
- ☐ *Ferdinandea buccata* (Loew, 1863)
- ☐ *Hiatomyia cyanescens* (Loew 1863) *
- ☐ *Rhingia nasica* Say, 1823
- ☐ *Copestylum caudatum* Curran 1927
- ☐ *Copestylum vittatum* Thompson 1976
- ☐ *Volucella evecta* Walker 1852
- ☐ *Volucella facialis* Williston 1882

SUBFAMILY MICRODONTINAE

- ☐ *Microdon abditus* Thompson 1981
- ☐ *Microdon globosus* (Fabricius 1805)
- ☐ *Microdon albicomatus* Novak 1977 *
- ☐ *Microdon cothurnatus* Bigot 1884 *
- ☐ *Microdon manitobensis* Curran 1924
- ☐ *Microdon megalogaster* Snow 1892
- ☐ *Microdon tristis* Loew 1864
- ☐ *Mixogaster breviventris* Kahl, 1897
- ☐ *Omegasyrphus coarctatus* Loew 1864

Syrphidae of Minnesota by Horace Telford (1932): Checklist and Synonyms

Asemosyrphus willingii = *Arctosyrphus willingii*

Baccha fascipennis = *Hypocritanus fascipennis*

Cartosyrphus leucoparea = *Cheilosia leucoparea*

Cartosyrphus tristis = *Orthonevra tristis*

Cerioides abbreviata = *Ceriana abbreviata*

Cheilosia angelica = *Cheilosia orilliaenis*

Cheilosia minnesotensis = *Cheilosia orilliaenis*

Chrysogaster nigripes = *Chrysogaster antitheus*

Chrysotoxum minor = *Chrysotoxum derivatum*

Citibaena tuberculatus = *Merodon equestris*

Cnemodon calcarata = *Heringia calcarata*

Cnemodon carinata = *Neocnemodon carinata*

Cnemodon coxalis = *Neocnemodon coxalis*

Cnemodon intermedia = *Neocnemodon intermedia*

Cnemodon pisticoides = *Neocnemodon pisticoides*

Condidea lata = *Sericomyia lata*

Cynorhina analis = *Blera analis*

Cynorhina confusa = *Blera confusa*

Elophilus fasciatus = *Helophilus fasciatus*

Elophilus hybridus = *Helophilus hybridus*

Elophilus latifrons = *Helophilus latifrons*

Elophilus obscurus = *Helophilus obscurus*

Epistrophe albipunctatus = *?*

Epistrophe cinctellus = *Meliscaeva cinctella*

Epistrophe fisherii = *Melangyna fisherii*

Epistrophe xanthostomus = *Epistrophe xanthostoma*

Eristalis bastardi = *Eristalis anthophorina*

Eristalis latifrons = *Eristalis stipator*

Eristalis nemorum = *Eristalis interrupta*

Ferdinandea dives = *Ferdinandea buccata*

Heliophilus angustiventris = *Xylota angustiventris*

Heliophilus chalybea = *Chalcosyrphus chalybeus*

Heliophilus curvipes satanica = *Chalcosyrphus satanica*

Heliophilus ejuncida = *Xylota ejuncida*

Heliophilus nemorum = *Chalcosyrphus nemorum*

Heliophilus pigra = *Chalcosyrphus piger*

Heliophilus vecors = *Chalcosyrphus vecors*

Heryngia salax = *Heringia salax*

Lathyrophthalmus aeneus = *Eristalinus aeneus*

Lejops bilinearis = *Anasimyia bilinearis*

Lejops chrysostomus = *Anasimyia chrysostomus*

Lejops lunulatus = *Anasimyia lunulatus*

Lejops relictus = *Anasimyia chrysostomus*

Lejops stipatus = *Eurimyia lineatus*

Mallota cimbiciformis = *Mallota bautias*

Mallota facialis = *Mallota bautias*

Melanostoma obscurum = *Platycheirus obscurus*

Melanostoma pictipes = *Platycheirus pictipes*

Mesogramma marginata = *Toxomerus marginatus*

Mesogramma polita = *Toxomerus politus*

Metasyrphus amalopis = *Dasysyrphus intrudens*

Metasyrphus canadensis = *Eupeodes americanus*

Metasyrphus divisus = *?*

Metasyrphus emarginatus = *Epistrophe emarginata*

Metasyrphus lapponicus = *Eupeodes lapponicus*

Metasyrphus laticaudus = *Dasysyrphus pauxillus*

Metasyrphus latifasciatus = *Eupeodes latifasciatus*

Metasyrphus medius = *Eupeodes americanus*

Metasyrphus neoperplexus = *Eupeodes neoperplexus ?*

Metasyrphus perplexus = *Eupeodes perplexus*

Metasyrphus weborgi = *Epistrophe emarginata*

Metasyrphus wiedemanni = *Eupeodes americanus*

Microdon baliopterus = *Omegasyrphus baliopterus*

Microdon pseudoglobosus = *Microdon globosus*

Microdon robusta = *Microdon tristis*

Orthoneura nitida = *Orthonevra nitida*

Orthoneura pictipennis = *Orthonevra pictipennis*

Paragus tibialis = *Paragus haemorrhous*

Pipizella banksi = *Trichopsomyia banksi*

Pipizella recedens = *Trichopsomyia recedens*

Platycheirus erraticus = Platycheirus hyperboreus

Pyrophaena granditarsis = Platycheirus granditarsis

Pyrophaena rosarum = Platycheirus rosarum

Sphaerophoria cylindrica = Sphaerophoria contigua

Sphaerophoria robusta = Sphaerophoria philanthus

Spilomyia quadrifasciata = Spilomyia sayi

Syrphus bigelowi = Syrphus ribesii

Syrphus ribesii vittafrons = Syrphus ribesii

Temnostoma apiforme = Temnostoma excentrica

Temnostoma trifasciata = Temnostoma trifasciatum

Toxomerus geminata = Toxomerus geminatus

Volucella bombylans americana = Volucella facialis

Volucella bombylans evecta = Volucella evecta

Volucella fasciata = Copestylum vittatum

Cedar Creek & University of Minnesota: Checklist and Synonyms

Microdon pseudoglobosus = Microdon globosus

Chrysogaster nitida = Chrysogaster antitheus

Cynorhina analis = Blera analis

Cynorhina confusa = Blera confusa

Xylota nemorum = Chalcosyrphus nemorum

Xylota pigra = Chalcosyrphus nigromaculatus

Xylota vecors = Chalcosyrphus vecors

Chrysotoxum minor = Chrysotoxum derivatum

Chrysotoxum cf fasciolatus = Chrysotoxum fasciolatum

Eristalis bastardi = Eristalis anthophorina

Eristalis burda = Eristalis flavipes

Eristalis nemorum = Eristalis interrupta

Eristalis latifrons = Eristalis stipator

Metasyrphus medius = Eupeodes americanus

Metasyrphus wiedemanni = Eupeodes americanus

Metasyrphus latifascialis = Eupeodes latifasciatus

Lejops bilinearis = Anasimyia bilinearis

Lejops lunulatus = Anasimyia lunulatus

Lejops stipatus = Eurimyia lineatus

Mallota cimbiciform = Mallota bautias

Epistrophe fischeri = Melangyna fisherii

Mixogaster sp. = Mixogaster breviventris

Neoascia globosa = Neoascia globosa

Orthonevra sp. A = Orthonevra nitida

Orthonevra sp. B = Orthonevra pulchella

Paragus tibialis = Paragus haemorrhous

Pyrophaena rosarum = Platycheirus rosarum

Condidea lata = Sericomyia lata

Sphecomyia quadrivittata = Sphecomyia vittata

Sphegina flavimana = Sphegina flavimana

Mesograpta marginata = Toxomerus marginatus

Xylota ejuncida = ?

iNaturalist and BugGuide: Synonyms

Lejops willingii = Arctosyrphus willingii

Lejops bilinearis = Anasimyia bilinearis

Lejops chrysostomus = Anasimyia chrysostomus

Lejops lunulatus = Anasimyia lunulatus

Lejops relictus = Anasimyia chrysostomus

Lejops stipatus = Eurimyia lineatus

Lejops curvipes = Polydontomyia curvipes

Ocyptamus fascipennis = Hypocritanus fascipennis

INDEX

INDEX

ABOUT THE AUTHOR

Scott King is the founder and editor of Red Dragonfly Press. He is the author of several books of poetry, most recently *All Graced in Green* and *Dragonfly Haiku* (w/ Ken Tennessen), and translator of books by Greek poet Yannis Ritsos and Persian poet Fereydoun Faryad. In recent years he has begun writing natural history, publishing a series of volumes of field notes about dragonflies and a guidebook to the wasps and bees of Minnesota.

In 2017, Scott King set himself the challenge of submitting at least one observation to an online, citizen science database each day of the year. These daily observations consisted of photos and a journal entry. The 365 journal entries were subsequently published as Following the *Earth Around: Journal of a Naturalist's Year* (Thistlewords Press, 2019).

He has numerous writing projects in progress but hopes to finish a revision of *Dragonflies and Damselflies of Minnesota: Atlas and Annotated Checklist* (Thistlewords Press, 2018) and publish his decade-long project *Meadowhawks: A Natural History of the Red Dragonflies*.

A lifetime resident of Minnesota, he currently resides in Northfield.

Minnesota Flower Flies
www.MinnesotaFlowerFlies.com

Of Books and Bugs
OfBooksAndBugs.wordpress.com

Meadowhawks
www.Meadowhawks.info